Aviation versus Environment?

AF784365

Aviation *versus* Environment?

2nd Hamburg Aviation Conference

edited by
Thomas Immelmann,
Otto G. Mayer,
Hans-Martin Niemeier,
Wilhelm Pfähler

PETER LANG
Europäischer Verlag der Wissenschaften

Die Deutsche Bibliothek - CIP-Einheitsaufnahme

Aviation versus environment? / Thomas Immelmann ... (eds.) . -
Frankfurt am Main ; Berlin ; Bern ; Bruxelles ; New York ;
Wien : Lang, 2000
ISBN 3-631-35874-1

Jacket design:
Flughafen Hamburg GmbH

ISBN 3-631-35874-1
US-ISBN 0-8204-4389-1

© Peter Lang GmbH
Europäischer Verlag der Wissenschaften
Frankfurt am Main 2000
All rights reserved.

All parts of this publication are protected by copyright. Any utilisation outside the strict limits of the copyright law, without the permission of the publisher, is forbidden and liable to prosecution. This applies in particular to reproductions, translations, microfilming, and storage and processing in electronic retrieval systems.

Printed in Germany 1 3 4 5 6 7

Preface

Continued exponential rates of growth in air transport in many parts of the world are resulting in significant environmental pressures in spite of technological advances in aircraft and engine design. Airport transportation is part of modern life and general welfare. Like in any other form of transportation, energy is used; for generating this energy, fuel is burned. By burning fuel various gases and numerous types of particles are produced and released into the atmosphere. Substantial research is being carried out worldwide to investigate the effects of aviation. This is a big task, not the least due to the complexity of the atmospheric system. It appears from all this work that under most circumstances the effects of aviation for the global atmosphere are modest and scaled to the actual amount of fuel burned higher up in the troposphere. There are no extreme sensitivities that have been discovered. One cannot, however, rule out surprises, as surprises are part of incomplete understanding and information.

Airport activities, of course, produce environmental problems, too, with noise pollution being probably the most prominent issue. Environmental damage may be reduced by various measures. For example, airports may invest in sound absorbing buildings either at the airport site or in the neighbourhood, they may introduce noise related landing fees or a general ban on certain types of aircraft or on night flights. All these measures either increase costs of airport operation or reduce the airport's revenue. On the other hand there are the social costs of environmental damage. Efficient use of infrastructure, therefore, requires that environmental costs are internalised. As the reduction of pollution is a public good, state intervention is required to get the right levels of polluting activities. In other words, in order to improve the general welfare economic policy must not only liberalize the aviation industry, but must also internalize the negative external effects of aviation.

Obviously, local and global externalities are closely linked. Modern aircraft and their engines are generally less noisy and have lower emission levels which are less damaging to the global atmosphere. If the use of modern aircraft can be fostered by the price policy of airport charges, then regional aviation policy contributes to diminishing problems of the global atmosphere. Likewise, airports and their regions would benefit from efficient measures fostering the use of modern aircraft, which, consequently, would reduce global emission levels. Such measures would reduce noise levels at airports and reduce the magnitude of local external effects.

This book sets out to lay ground for a solid analysis of the problems and policy options at hand, and bring together the best expertise available in this field.

Thus, renowned experts from Germany, Europe and overseas explore in depth the economics of air transport environmental measures, the possible atmospheric changes resulting from air transportation, the benefits and cost of alternative noise management strategies, the specific approaches of different airports and their experiences with their various environment protection measures.

Essentially this book contains the proceedings of the international conference on "Aviation versus Environment?", held in Hamburg on February 17-19, 1999, and jointly organised by the University of Hamburg, the HWWA Institute for Economic Research, Hamburg, the State Ministry of Economic Affairs of the Free and Hanseatic City of Hamburg and the Hamburg Airport. The authors have agreed to extend and revise their conference papers for the purpose of this publication. The editors are more than grateful for their participation at the conference and for their additional effort in making this book possible.

This conference was the second one in the newly created regular international conference series on air traffic and airports, held in Hamburg and named "Hamburger Luftverkehrskonferenz/ Hamburg Aviation Conference". The papers of the first conference in 1998 were published (Peter Lang GmbH, Frankfurt am Main, 1999) under the title "Airports and Air Traffic. Regulation, Privatisation, and Competition". Both conferences have been clearly a success and confirmed the organisers in their original intentions. The follow-up conference is scheduled to take place in February 2000 at the Hamburg Airport.

Thomas Immelmann
Hamburg Airport
Hans- Martin Niemeier
State Ministry of Economics, Hamburg

Otto G. Mayer
HWWA
Wilhelm Pfähler
University of Hamburg

Contents

Preface 5

Part I: Global Air Transport and Ecological Limits 9

Air Transport Environmental Measures:
The Economic Perspective
Peter S. Morrell, Cranfield University, U.K. 11

The Perspective of Natural Science- The CARIBIC project
(Civil Aircraft for Remote Sensing and In-Situ-Measurement
of Troposphere and Lower Stratosphere)
Carl A.M. Brenninkmeijer, Max-Planck-Institute for Chemistry,
Atmospheric Division, University of Mainz, Germany 25

Part II: International and European Environmental Policy: Protection of Industry or Environment? 37

The International Perspective
Upali K. Wickrama, ICAO International Civil Aviation Organization 39

Towards an efficient market-based environmental aviation policy –
A view from the Hamburg Region
Hans-Martin Niemeier, Ministry of Economics, Hamburg, Germany 47

Part III: Environmental Policy of Airports in the Age of Privatization 69

Benefits and Costs of Alternative Noise Management Strategies
David Gillen, University of California, USA 71

Achieving the Correct Balance – the Manchester Airport Second
Runway Project
Callum Thomas, Manchester Metropolitan University, UK 111

Hamburg Airport – Environmental Management System
Axel Schmidt, Environmental Affairs Dept., Hamburg, Germany 123

Airport Zurich – Charging Polluting Emissions
Emanuel Fleuti, Zurich Airport, Switzerland 129

Short Biographies 131

PART I

GLOBAL AIR TRANSPORT AND ECOLOGICAL LIMITS

Air Transport Environmental Measures: The Economic Perspective

Peter Morrell
(Air Transport Group, College of Aeronautics, Cranfield University, UK)

Introduction

This paper discusses the relationship between economic development, air transport growth and its effect on the environment. In particular, it focuses on approaches to controlling polluters and compensating those affected by pollution (the externality problem). Both short and long-term measures for reducing pollution from aircraft are included, as are the growing number of aviation related environmental charges.

Continued exponential rates of growth in air transport in many parts of the world are resulting in significant environmental pressures, in spite of technological advances in aircraft and engine design. Some signs of market maturity might be observed, although the evidence from the US and UK suggests a longer term slowing down, rather than a levelling off, of growth (Figures 1 and 2). It can also be argued that environmental taxes and other constraints which force up prices will hasten the onset of maturity in major world markets.

Figure 1

Total UK Airport Passenger Traffic

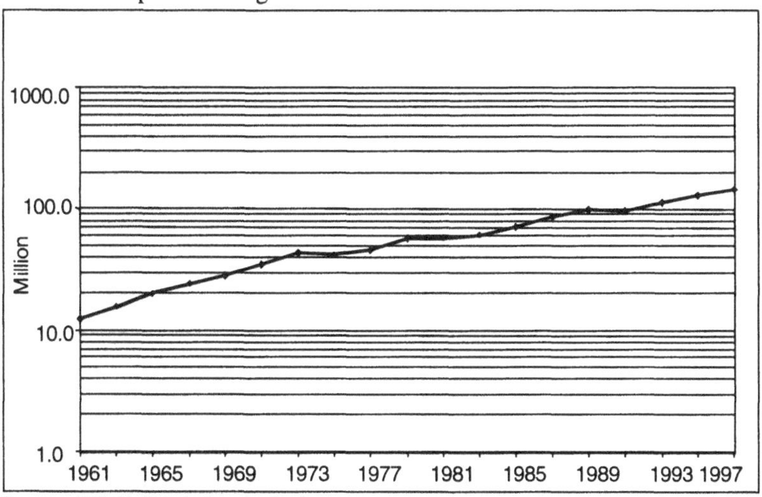

Source: UK Civil Aviation Authority

European air transport liberalisation has played a key role in the resumption of higher growth rates from UK (and European) airports since 1993. It has been argued that European liberalisation was important in generating both higher numbers of passengers and a greater number of aircraft movements than would have been the case under the previous bilateral regime (Cranfield, 1997). This was accompanied by a downsizing of aircraft, in order to increase frequency to gain competitive advantage. However, the environmental impact of such changes are far from clear, especially where fuel-efficient quieter smaller aircraft (eg Airbus A320s) are replacing older technology larger aircraft such as Airbus A300s.

In the US, some levelling out of the passenger growth since the mid to late 1980s is evident from Figure 2. This may have been caused by the failure of many low cost start-up airlines and the consolidation of the industry (leading to higher air fares), rather than to any sign of market maturity.

Figure 2

Total US Enplanements

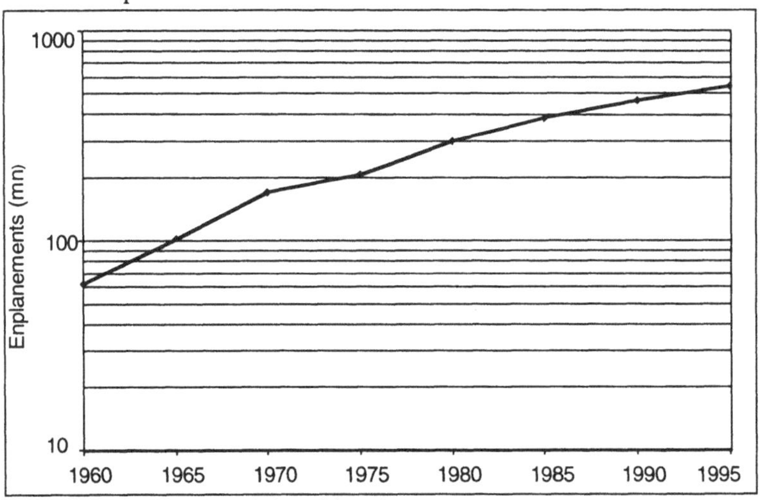

Source: Air Transport Association of America

Even without the increasing incident of environmental taxes, which are discussed more fully later, technological developments such as video-conferencing, and, perhaps in the longer term, virtual reality holidays can provide realistic alternatives for some consumers of air travel. The success of these, and the timescale for their introduction, will depend on longer term socio-economic factors.

Increasing consumer awareness of the environmental impact of air travel may also lead to more eco-friendly holidays being offered. So far, these have been confined to limiting the ground impact of tourism (eg waste disposal), although many airlines now produce environmental reports showing their commitment to improving aircraft noise and emissions, as well as waste management and congestion (British Airways, 1998).

The growing incidence of airport charges and governmental taxes on aviation could also lead to slower growth. The UK government introduced a tax on air travel: the non-EU destination rate was originally £10 per departing passenger, but this was later increased to £20. This suggested that the UK Treasury considered that the overall market price elasticity was not sufficiently high for the tax to cause any significant reduction in air travel. A similar conclusion could be drawn from the majority of environmental charges discussed later in this paper.

Finally the elimination of government subsidies to airlines, particularly in Europe, should have been welcomed by environmentalists. This has resulted in more market based pricing, and the elimination of some uneconomical routes being served. Airport privatisation and slot trading are further measures aimed at introducing more market disciplines to air transport, with potential environmental benefits in terms of less congestion and more community focused policies (see Doganis, 1999 and Wolf, 1999).

1.0 Externalities – environmental costs

Given the aim of sustainable development, which seeks continued economic growth with limited environmental deterioration, more and more airports consider environmental constraints to be amongst the most important factors which determine airport capacity (OECD, 1997; Keur, 1998).

Amongst the externalities generated from commercial flights, noise nuisance has the greatest impact on the community surrounding the airports; on the other hand, engine emissions during the cruise mode are the only direct man-made source of exhaust pollutants within the upper troposphere and lower stratosphere. Furthermore, due to the serious delays that many major airlines are facing, the amount of exhaust pollutants, both in the air and on the ground, are increasing significantly.

Of the various environmental instruments, environmental charges have been proved to be an effective and economic way to internalise the externalities, while, on the other hand, encouraging airlines to use more environmentally friendly aircraft (OECD, 1994; EEA, 1996). However, given the fact that airlines cannot make significant changes to their fleet in the short to medium term, the environmental measures and restrictions are also essential for mitigating the emissions level.

2.0 International aircraft noise standards: ICAO Annex 16

The first and most obvious environmental classification of aircraft is the one described in ICAO's Annex 16 noise standards (ICAO, 1991), where jet aircraft are subdivided into 3 different Chapters according to their noise certification:

- Chapter 1: aircraft with type licence before 1970 (e.g. B707s)
- Chapter 2: aircraft with type licence between 1970 and 1978 (e.g. B747-100/200s, certain B737-200s)
- Chapter 3: aircraft with type licence after 1978 (e.g. A310, B757, B767, MD83)

An international standard of noise compliance was agreed through ICAO, such that "all Chapter 2 aircraft would be banned from 1st April 2002," and accepted by the airlines worldwide. However, the European Commission tends to apply a more stringent regulation against hushkitted Chapter 2 aircraft operating within EU nations, which has recently been the subject of much controversy between USA and EU.

3.0 Noise measures at selected European airports

The purpose of this section is to describe and assess, in terms of broad economic impact, the current airport environmental management or abatement measures on aircraft noise. The examples of selected European airports are analysed later with respect to noise abatement procedures as well as noise surcharges/discounts and penalties.

Short questionnaires were sent to most of the survey airports so as to ensure the information used in this research was accurate and up-to-date. Where no reply was obtained, published sources were used, as well as airline flight operations departments, to check whether any significant environmental measures were in effect (ACI Europe, 1999; Boeing, 1999; IATA, 1999). The results are shown in Table 1. The environmental management measures are generally very specific to particular airports. They also vary considerably in both the effect they have on noise nuisance, and the cost impact on airlines.

Table 1

Noise related restrictions for survey airports

Country	Airport	Noise Abatement Procedures[1]	Noise Related Charges[2]	Noise Penalties[3]
France	Paris-CDG	Yes	Yes	Yes
Germany	Hamburg	Yes	Yes	Yes
	Munich	Yes	Yes	Yes
Italy	Rome-Fiumicino	Yes	Yes	Yes
Netherlands	Amsterdam-Schiphol	Yes	Yes	No
Portugal	Lisbon	Yes	No	No
Spain	Madrid	Yes	No	Yes
Sweden	Stockholm-Arlanda	Yes	Yes	No
Switzerland	Zurich	Yes	Yes	No
United Kingdom	London-Heathrow[4]	Yes	Yes	Yes

Source: Direct airport contacts; ACI Europe, 1999; Boeing, 1999

Note:

1. Noise abatement procedures mean the regulations on aircraft landing and take-off procedures, as well as on flight paths.
2. Noise related charges are all forms of charges which take the aircraft acoustic level into consideration.
3. Noise penalties mean the fines on aircraft deviating from the flight paths as well as exceeding the regulated maximum noise level.
4. London Gatwick and London Stansted airports have the same noise measures as London Heathrow Airport.

3.1 Noise abatement procedures

It can be seen from Table 1 that landing and take-off noise abatement procedures are applied at all survey airports. In all cases, these take the form of procedures for take-off and landing, and in most cases there are also runways designated either for take-offs or landings; in only a few cases are there limits on the maximum noise levels allowed.

3.1.1 Low noise approach paths

Low noise approach paths usually take the form of a steeper descent during the first phase of the approach (5-6°), combined with a normal second phase (3°). The advantage of this two segment approach is that noise is reduced, say, at around 3-5 nautical miles from the runway threshold, with the effect of higher altitude outweighing the greater noise from increased flap settings. Estimates of noise reduction using this procedure in the US ranged from 10 EPNdB at 5 nm to 6 EPNdB at 3 nm.

Other procedures are low drag approaches with reduced flap settings, but this is not thought to give a very significant improvement. Continuous descent approaches also generate less noise from lower power settings, while at the same time being more economic for operators.

Intercepting the ILS glide-slope further out and at a higher altitude will also give some relief to communities living closer to the airport, but may increase the noise impact somewhat on those living slightly further out. On balance, however, there might be a small net gain.

3.1.2 Later lowering of landing gear

Procedures for lowering the landing gear closer to the runway threshold would also reduce noise somewhat. This has no significant impact on airline operating costs, but probably also has little impact in reducing noise nuisance.

3.1.3 Ban on reverse thrust use

The use of reverse thrust on landing generates about 10 dB less noise than produced by the engines operated normally during take-off mode, but the noise is sudden and could cause nuisance to surrounding communities, especially at night. Thus a night ban on the use of this was found to be quite common at EU airports.

The impact of this on airline costs was thought to be relatively neutral: increased costs of maintaining brakes is balanced by the reduced costs of engine maintenance. In terms of safety, however, some airlines indicated that safety margins were increased using reverse thrust, with brakes still available as back-up. The use of brakes only, with reverse thrust as a back-up is not as effective an option, since the seconds lost in getting engines from idle to full power would be critical in any emergency.

3.1.4 Noise abatement take-off procedures

Noise can be reduced by cutting back power, once the aircraft has reached a safe operating altitude. Reduced power is then used until the aircraft is no longer over a populated area, when full power can be resumed. This procedure may result in some overall reduction in the noise impact, and will also re-distribute noise away from those living nearer the runway threshold to communities further away from the airport, who will be affected by aircraft flying at a lower altitude.

Low noise departures can also be made by increasing the angle of ascent. But there is a trade-off between the lower speed that is used in connection with the higher rate of climb, and the higher speed associated with the lower rate of climb. The same thrust is used in both cases. The first procedure imposes a higher noise nuisance over a shorter period of time, while the second a lower noise over a longer period of time. The precise impact will depend on the noise impact measure used, but this suggests that the net improvement may not be very large.

Noise preferential routes are also used at many EU airports, designed to minimise noise exposure to more densely populated communities. Short/medium haul aircraft tend to be more manoeuvrable, and thus find these routings easier to comply with, compared to heavier long-haul aircraft such as B747s. The cost implications of such routes are difficult to estimate, since they would sometimes mean a greater distance flown and at other times a shorter distance.

3.1.5 Runway operational modal split (noise preferential runways)

A number of European airports designate particular runways for either take-off or landings, so as to minimise the noise exposure experienced by more densely populated communities. The main examples of this in Europe are:

- Amsterdam (one runway used only for take-offs)
- Frankfurt (one runway used only for take-offs)
- London Heathrow

At Heathrow, each of the two parallel runways is operated in either take-off or landing mode, so as to reduce the continuous exposure to both take-off and landing noise (segregated mode operations). The runway mode is switched every 12 hours.

These measures do not have any significant direct cost impact on airlines, but they do reduce the effective number of peak hour movements. Since some of these airports are also slot co-ordinated, this limits expansion and increased frequency possibilities, and also leads to greater congestion.

3.2 Noise surcharges and discounts

Airports in eleven European and four Asian countries, as well as the United States, currently apply aircraft noise surcharges and discounts. Most airports apply a percentage surcharge or discount on the MTOW based landing fee, depending on the aircraft acoustic category (the Swiss airports apply a fixed amount surcharge).

In the case of German and BAA London airports, the total landing fee varies according to aircraft acoustic noise category, such that it is impossible to separate out the noise element of the charge (Morrell and Lu, 1999).

Tables 2 and 3 show the noise surcharge or reduction on B767-300 Chapter 3 aircraft and B737-200 Chapter 2 aircraft at selected five airports where this can be shown separately, expressed in terms of Euros per seat.

The surcharges for the B767-300 Chapter 3 aircraft revealed some inconsistencies: The charges at Amsterdam Airport Schiphol were comparatively higher than those at other airports. However, this aircraft type received a discount at Paris CDG and Stockholm-Arlanda. There was no noise surcharge on B767-300 at Zurich airport, as this aircraft type is classified in the quietest aircraft acoustic category. Of all the five airports, three airports, Paris-CDG, Rome-Fiumicino and Amsterdam-Schiphol airports, applied night-time surcharge on both Chapter 3 and Chapter 2 aircraft.

Table 2

Noise surcharges at selected airports – B767-300 (Chapter 3 aircraft)

Unit: ? /seat

Country	Airport	Day-time	Night-time
France	Paris-CDG[1]	-0.64	-0.36
Italy	Rome-Fiumicino	0.08	0.86
Netherlands	Amsterdam-Schiphol[2]	0.74	1.30
Sweden	Stockholm-Arlanda	-0.16	-0.16
Switzerland	Zurich	0.00	0.00

Source: Direct airport contacts; IATA, 1999

Note:

1. The noise surcharge at Paris-CDG airport includes the "Noise Tax."
2. The noise surcharge at Amsterdam Airport Schiphol includes the "Governmental Noise Charge," imposed by the Dutch government at the majority of the Dutch airports.

Table 3 shows that there was a large range of surcharges applied for a day-time landing with a Chapter 2 B737-200, ranging from only ? 0.13 at Rome to ? 13.78 at Amsterdam Airport Schiphol.

Table 3

Noise surcharges at selected airports – B737-200 (Chapter 2 aircraft)

Unit: ? /seat

Country	Airport	Day-time	Night-time
France	Paris-CDG[1]	2.45	7.90
Italy	Rome-Fiumicino	0.13	0.45
Netherlands	Amsterdam-Schiphol[2]	13.79	14.08
Sweden	Stockholm-Arlanda	2.38	2.38
Switzerland	Zurich	1.92	1.92

Source: Direct airport contacts; IATA, 1999

Note:
1. The noise surcharge at Paris-CDG airport includes the "Noise Tax."
2. The noise surcharge at Amsterdam Airport Schiphol includes the "Governmental Noise Charge," imposed by the Dutch government at the majority of the Dutch airports.

However, apart from the surcharge on Chapter 2 aircraft at Amsterdam Airport Schiphol, the importance of these charges at current levels is very small on a per seat or even per passenger basis.

Table 4 shows how the noise surcharge revenues are used in five European countries. It can be seen that in all cases the airport noise charges were revenue neutral (or, in other words, they resulted in no net increase in revenues). Thus, increases in charges for noisier aircraft were balanced by reductions in charges for quieter aircraft. Where the number of Chapter 2 aircraft operations are declining sharply, discounts for quieter aircraft need to be reduced to maintain income. Of the five countries, four had noise abatement investment programmes, but these were funded from general revenues. The situation for Amsterdam Schiphol is more complicated, in that both airport and government levy noise surcharges, the revenue from which goes toward the noise insulation scheme (managed by the airport). While the airport scheme is revenue neutral, the government aircraft noise tax is not, but is ring-fenced for airport noise abatement investments (Dutch Aeronautical Inspection Directorate, 1998).

Table 4

The application of noise surcharges at selected airports

Country	Airport	No increase in revenue	Noise abatement investment	Pass on to government
Germany	Majority of airports	✓	✓	
Italy	Rome-Fiumicino	✓		✓
Netherlands	Amsterdam-Schiphol*	✓	✓	✓
Sweden	Stockholm-Arlanda	✓	✓	
Switzer-land	Zurich	✓	✓	

Note:

- The money collected from the "Governmental Noise Charge" imposed by the Dutch government is entirely used for the soundproofing projects around the airports.

4.0 Conclusions

There is a growing impatience with the speed of reduction in environmental social costs in air transport. This is because of the difficulty in tightening international regulations, such as those on aircraft noise agreed through ICAO. Thus regional and local initiatives have been taken, especially in Europe. A combination of international regulations and standards with local (airport) measures and incentives seems to work well for aircraft noise. For example, Amsterdam Schiphol Airport has eliminated operations with noisier Chapter 2 aircraft through the application of taxes, prior to the internationally agreed deadline of 2002. However, pollution from engine emissions needs to be approached at the global level, with airport measures confined to dealing with relatively small local effects.

There is a trend of gearing fines and penalties more towards the actual noise incurred, through increased noise measuring and flight tracking systems being installed. However, the majority of noise surcharges have little economic rationale, and are unlikely to bear much relationship to actual environmental social costs imposed.

Pressures on land around airports and near large cities have, at some airports, reduced the effectiveness of zoning policies. These were designed to prevent the area surrounding airports being occupied by schools, hospitals, offices and private housing. The absence of such measures has increased the need for noise insulation schemes and night curfews, although some airports had been built close to city centres before zoning had even been considered.

Finally, there is little sign of long-term relief of pressures on the environment from significantly slower traffic growth rates, although economic recessions have a role in short term.

References

ACI Europe, http://www.aci-europe.org/, Airport Council International, Accessed February 1999.

Boeing, http://www.boeing.com/assocproducts/noise/airports.html, Airport Noise Regulation Information, Accessed February 1999.

British Airways, *Annual Environmental Report*, July 1998.

Cranfield University (Air Transport Group), *The Single Market Review, Subseries II: Impact on Services, Volume 2: Air Transport*, Kogan Page for the Office for Official Publications of the European Communities, 1997.

Doganis, R, *The Future Shape of the Airline and Airport Industries*, in Airports and Air Traffic, eds. Pfähler, W., Niemeier H-M, and Mayer, O, Peter Lang, 1999.

EEA, *Environmental Taxes: Implementation and Environmental Effectiveness*, European Environment Agency, Copenhagen, Denmark, 1996.

Dutch Aeronautical Inspection Directorate, *Governmental Noise Charges on Dutch Airports*, Aeronautical Inspection Directorate, Airworthiness Department, Noise Certification and Noise Charges, the Netherlands, November 1998.

IATA, *IATA Airport and En-route Aviation Charges Manual*, February 1999.

ICAO, *International Standards and Recommended Practices, Environmental Protection, Annex 16 to the Convention on International Civil aviation, Vol. 1, Aircraft Noise*, International Civil Aviation Organisation, Montreal, Canada, 1991.

Keur, J., *Minimise the Impact of Noise Limitations on Day-to-day Business and Profitability*, IIR's conference: Practicalities of Environmental Management in the aviation Industry, London, UK, 16th & 17th November 1998.

Morrell, P. and Lu, H., *Current Environmental Management Measures in Air Transport*, Aerogram, Vol. 9, No. 3, June 1999, pp. 9-14.

OECD, *Internalising the Social Costs of Transport*, ECMT, Paris, France, 1994.

OECD, *Environmental Taxes and Green Tax Reform*, OP-OE/3667, Paris, France, 1997.

Wolf, H, *Tackling Congestion and Environmental Problems*, in Airports and Air Traffic, eds. Pfähler, W., Niemeier H-M, and Mayer, O, Peter Lang, 1999.

Aviation and atmospheric change

Carl A. M. Brenninkmeijer
(Max Planck Institute for Chemistry)
Atmospheric Chemistry Division
Mainz

Introduction

Air transportation is part of modern life and welfare. Like in any other form of transportation, energy is used. For generating this energy, fuel is burned. By burning fuel, various gases and types of particles are produced and released into the atmosphere. In this overview we will look at the effects of these gases and particle emissions on the global atmosphere and environment.

Substantial research is being carried out worldwide to investigate the effects of aviation. This is a big task, not the least due to the complexity of the atmospheric system. Aviation, like many of our other activities, e.g. road traffic, deforestation, industrial production, agriculture, etcetera, change the global atmosphere. This occurs under our very own eyes, even before we have had a chance to completely understand the atmosphere and its chemistry in detail.

Atmospheric research has been fuelled by awareness of atmospheric changes. Some changes are, or may be, extremely nasty, like the spring time ozone hole over Antarctica. We will have to live with its consequences for decades to come. At the moment of writing, the Intergovernmental Panel on Climatic Change (IPCC) is tabling its report about the environmental effects of air traffic. I refer to this report for details. Here I give a bird's eye view of the present situation, and attempt to show that many processes can be understood qualitatively relatively easily. In addition to this I will show some excellent examples of progress in atmospheric research supported by the aviation industry.

Which atmospheric changes are of concern?

Before looking in detail at the process level, we emphasize the focus of this brief paper, i.e.; "Do aircraft emissions change the global atmosphere to an important degree? Followed by "Does this change have undesirable consequences, and which are these?" At the outset we acknowledge that there are, to our best knowledge, extremely serious non-aircraft-specific changes. These are part of the bigger picture, i.e. us burning fossil fuels to live the way we do. In doing this, the greenhouse gas carbon dioxide is emitted in vast quantities. By this we very likely enhance global warming, the greenhouse effect, or better formulated induce "climatic change". The earth's climate system and its variability are not well understood yet. Major scientific questions have remained unanswered up to date. Notwithstanding, the consensus is that the increase in greenhouse gases is an extremely serious concern. The scientific framework that forms the basis for this is well researched and some trust in science is not misplaced.

Of aviation specific concern are changes in ozone. Ozone filters out most of the harmful ultraviolet light from the sun. Higher in the atmosphere ozone levels seem to have been declining by which ultraviolet radiation increases. This increase is of concern for the wellbeing of man, animal, and plants. In other parts of the atmosphere ozone is increasing, and at ground level this may affect plants. In large parts of the remote atmosphere however, ozone also plays a role as a greenhouse gas, thus it contributes to the warming at the earth's surface. A third role of ozone is that it participates in the oxidizing (read "self cleansing") capacity of the atmosphere. Less ozone may lead to a reduction in the capacity of the atmosphere to rid itself of natural and man made pollutants. This in turn may enhance the greenhouse effect. For instance, the greenhouse gas methane may increase somewhat more. It is clear that these are all complex and coupled processes.

Of further concern are not only the gases in the atmosphere, but also changes in the quantity of fine particles and in the related changes in cloud cover. Small particles can enhance cloud formation, and clouds change the radiation balance of the atmosphere. Clouds reflect light effectively but also may absorb light, and thin high clouds keep the earth's surface somewhat warmer. Much uncertainty is associated with this set processes.

Aircraft, their engines, fuel, and exhaust products

Every single day millions of passengers and goods are safely transported at high speed and high altitude. Speed and altitude go hand in hand. Only in thin air, higher up, high speeds can be reached. Most of us do not realize the engineering problems of pushing trains at high speed through the denser air at ground level. Modern jets reach at the impressive speed of 250 meter per second with an astonishingly good fuel efficiency per passenger kilometer. This efficiency is the result of excellent aerodynamics combined, with highly efficient, sophisticated, jet engines.

The 2, 3, or 4 jet engines, which often seem rather small attachments to the aircraft, burn tons of aircraft fuel at extreme high temperatures. Temperature and efficiency go hand in hand. The thin air is compressed, strongly heated by burning fuel, and rapidly expands to propel the aircraft.

The hot gases spewed out by the jet engine form a mixture that fascinates certain engineers and chemists alike, but may worry us somewhat.

The exhaust of the jet engine consists of air, which has lost oxygen needed for burning the fuel, mixed with the various waste products. These products are mainly carbon dioxide (CO_2, the well know greenhouse gas) and water (H_2O, which often forms the bright visible trails of billions of tiny ice crystals). A very small fraction of the fuel is not completely burned. The corresponding products are mainly particulate carbon (C, like chimney soot), carbon monoxide (CO, a poisonous gas) and hydrocarbons (fuel fragments).

Jet engine fuel also contains small fractions of sulfur compounds. This sulfur appears as sulfur dioxide (SO_2, burned sulfur, in the past used as a disinfectant), and actually plays an interesting role in particle formation (droplets of H_2SO_4, sulfuric acid, like the acid in your car battery). Furthermore jet fuel contains traces of other chemicals including additives, like car petrol, which ultimately form a range of ultra low traces of air pollution, which we will not study here.

Although the hot exhaust gases initially do not contain many small or larger particles, the air in the trail of an aircraft is strongly modified and shows a variety of condensation and particle formation processes. A recent discovery is the large role that charged particles can play. Apparently, the chemical reactions in the engine are responsible for the formation of many ions, which can promote the subsequent formation and growth of particles in a variety of ways hitherto little researched.

Jet engines, being based on high temperature burning do produce nitrogen monoxide (simply NO) in significant quantities, like all other combustion processes do. This gas NO is not a product of the fuel. It is a product of air itself. Oxygen (O_2) and nitrogen (N_2), which are the main components of air, combine under the high temperatures to form this chemically active gas, NO. There is much discussion about this gas, because it forms such an important chemical reaction partner of the much debated ozone (O_3). Table 1 shows a compilation of most of the compounds discussed so far.

Table 1

Pollutant	Comments
CO_2	Combustion product, a major greenhouse gas.
(O_2)	Oxygen is consumed by the combustion, but this in itself is of no concern.
NO	Nitrogen monoxide, an active catalyst affecting ozone.
CO	Carbon monoxide. Product of incomplete combustion, reactive.
Hydrocarbons	Fuel fragments. Products of incomplete combustion, reactive.
SO_2	Sulfur dioxide. Depends on the sulfur content of the fuel. Forms particles.
Particles	Very fine particles or their initial nuclei are formed.
H_2O	Combustion product. Is a greenhouse gas. Also forms contrails.
Contrails	Long traces of ice crystals, form under certain conditions by mixing.

The atmosphere and its structure

For understanding potentially important effects of aviation on the atmosphere, we condense its chemistry and structure in a nutshell. The atmosphere is a thin layer of air that clings to our planet. When you go up, the air becomes thinner, dryer and colder. While flying you will have noted the typical uncomfortable outside temperatures of minus 40 to 60 degrees displayed on the flight information screen.

The upper layer of the atmosphere is called "stratosphere" ("stratified", or "layered"). In this stratosphere, ultraviolet radiation of the sun is absorbed by ozone (O_3). This has 2 enormous important consequences. One; the stratosphere is actually warmer, and the warmer air floats on the colder air underneath. Two; the stratospheric ozone layer held in place by the layered structure protects life on earth for the harmful ultraviolet radiation.

The lower layer of the atmosphere is the troposphere. This layer is not stratified but continuously mixed in enormous turbulent processes. A lot of chemistry takes place here, again as part of natural processes. In the gas phase, and inside rain water droplets, and also on the surfaces of particles, important, often complex chemical and physical reactions take place.

The region between the stratosphere and the troposphere is called the tropopause. The tropopause is a sort of boundary with a variable height above the earth's surface. At the poles, it is at its lowest at 7 (winter) to 10 km (summer). In the tropics, the tropopause is lofted to as high as 18 km due to the intense heating of the air at the surface in equatorial regions.

The tropopause is a very complex boundary and the transport of gases and particles across this boundary is not documented in great detail yet.

As it happens, air traffic occurs mostly in this complex tropopause region. This means that we are faced with a complex situation with many unknowns. Added to this, it is expensive to do measurements with aircraft, and, the tropopause region, which we often observe while quietly flying, really is "aero incognita".

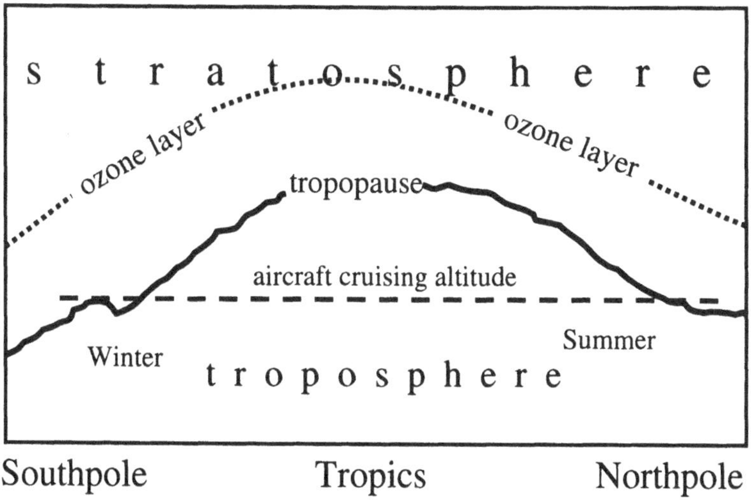

This figure shows the main structure of the atmosphere. Displayed is the situation during the summer in the northern hemisphere. The aircraft cruising altitude is between 9 and 12 km, thus usually below the tropopause boundary. Note that this boundary is neither sharp, nor static. Below it is the troposphere, which is well mixed and contains clouds. Above it is the dry, stratified stratosphere. For certain routes an aircraft will cross the tropopause boundary and enter the stratosphere. Note that in winter the tropopause is lower than in summer. In the tropics no emissions into the stratosphere will take place. The tropopause level here is too high for the present fleet of subsonic aircraft. The ozone layer is not a sharp layer, but rather a diffuse region where ozone levels are strongly elevated. If all ozone in the stratosphere were to be brought to the surface of the earth, it would form a layer of pure ozone gas, only 3 mm thick. This is a very thin layer, but ozone protects us very effectively. Also note that in the picture the vertical scale is grossly exaggerated. In reality, the atmosphere is best compared with thin a sheet of paper…all processes of weather, chemistry, and climate take place in this thin, incessantly changing layer, in contact with the oceans, ice sheets, and land masses.

The atmosphere and its chemistry

The main components of the atmosphere, oxygen and nitrogen are relatively uninteresting because of their low chemical reactivity. Only at high temperatures, like in fires do reactions take place. However, chemically active gases are injected into the atmosphere on a colossal scale. We are talking about nearly halve a billion ton of methane (CH_4) for instance, which is largely a product of nature. Presently methane is increasing due to our activities on earth, like keeping large numbers of cattle and growing rice on vast scales. A balance is disturbed. The dynamic balance is that between the input of gases into the atmosphere, and their removal. This removal mechanism, forming the self-cleansing oxidative capacity of the global atmosphere, is closely linked to the presence of ozone, a fact discovered around 1970.

How does it work? A healthy atmosphere needs ozone, and also produces ozone. Ozone is imported from the stratosphere into the troposphere. How much ozone that actually is, forms a research topic for many scientific studies. This ozone is split by the left-over ultraviolet radiation penetrating into the troposphere to form extremely reactive oxygen radicals. Some of these react with water vapor, and thus form 2 hydroxyl radicals (OH). This hydroxyl actually forms the strong natural bleaching agent of the troposphere.

Hydroxyl attacks many pollutants (both natural and man made). A methane molecule survives on average for about 9 years; this is a tough molecule. A carbon monoxide molecule survives several months. Terpene molecules, emitted on a large scale by plants, perish in a matter of hours. When more pollutants are introduced into the atmosphere, hydroxyl levels change, and the balance is shifted. We see that the tranquil atmosphere is the scene of active photo chemistry, which is driven by sunlight and also needs the well known ingredients ozone and water vapor.

A very interesting and important role is played by nitrogen monoxide, NO. This role can be easily depicted and is the basis for the ozone smog formed in summer.

Reaction 1. **O_3 + ultraviolet light ⇒ O_2 + O* (Ozone is split, O* has energy)**

Reaction 2. **O* + H_2O ⇒ 2 OH (A water molecule is split to form 2 hydroxyls)**

Reaction 3. **OH + CO + O_2 ⇒ CO_2 + HO_2 (The hydroxyl oxidizes CO)**

Reaction 4. **HO_2 + NO ⇒ NO_2 + OH (Thanks to NO, OH is formed back)**

Reaction 5. **NO_2 + light ⇒ NO + O (NO_2 is actually easily split)**

Reaction 6. **O + O_2 ⇒ O_3 (Finally we have ozone back again)**

To make a long story short, and do some bookkeeping, which some of us like more than chemistry anyway, you will see that the overall miraculous effect of reactions 1 to 6 is that NO helps to form extra ozone. This continues many times until it is eventually itself removed from the scene. This catalytic behavior is typical for several important chemical reactions in our atmosphere. An awesome demonstration of such catalytic power is the destruction of millions of square kilometers of ozone layer (ozone hole formation) by the chlorine from spray cans, like the early deodorant propellant, and other industrial gases. Please appreciate that the processes leading to the formation of the spring time ozone over Antarctica are well understood, and that this understanding is actually a proof of our vastly improved understanding of the complex atmospheric chemistry. There is no doubt that we have caused the ozone hole; at least we understand it.

The question is raised whether the effect of NO is eventually the production of ozone, or possibly the destruction of ozone. This is one of the questions of the impact of aviation on the atmosphere. For very high flying aircraft NO is released in the stratosphere and this will lead to ozone depletion. However, we will not deal here with a scenario of a future fleet of high flying super sonic aircraft. For lower heights ozone formation is more likely. It is a well known fact that in industrialized, densely populated areas, ozone is formed by NO from cars and industry. This is part of the problematic ozone smog formation. In the tropopause region where aircraft cruise it is not so clear however.... Besides this, large amounts of NO are already generated by lightning, and this certainly is a widespread natural process.

Assessment of the effects of aircraft emissions

Let us first consider the emissions of the major greenhouse gas carbon dioxide. This immediately gives a direct measure of how to compare aircraft with other forms of transportation. Of the total of carbon dioxide emitted by the transportation sector, about 12 % is due to aircraft. This is a sizable fraction, and the climatic impact can be directly scaled to this percentage. It does not matter where the aircraft fly, the carbon dioxide is eventually evenly distributed through the whole atmosphere.

From the preceding sections it follows that NO emission are potentially very important, because of the chemical activity of NO. When we first consider the upper troposphere, NO from aircraft has to be gauged against existing background levels of this gas. On a global scale, aircraft NO is still a small source of this gas. There now is growing information about background NO levels. One major natural source is lightning, which in particular over the continents and in the tropics is ubiquitous. Further, over land, much NO is found at cruising altitudes that has been transported upwards by convection (the lofting of air). The origin of this NO is at the surface in areas of industrialization and concentration of population, caused by the intensive fossil fuel combustion. Another source of NO is biomass burning. The plumes drifting away from forest and savanna fires initially contain large amounts of NO and its oxidation products. Thus except for flight corridors, where aircraft NO can be fairly concentrated, aircraft have a small impact on existing upper tropospheric NO levels.

The effect of the additional NO from aircraft is a slight increase in tropospheric ozone. Over the mid northern hemisphere (mainly due to the intense air traffic in and between the US, Europe and Japan) there probably is about 5 % extra ozone due to aircraft. Basically the effect from the extra NO is similar to that of NO at ground level, causing in extreme cases the ozone smog. However, at the cruising altitude, ozone is not a health hazard, but active in the chemistry.

Altogether, it is expected that the greenhouse effect is very slightly augmented by this extra ozone. A coincidental benefit however, is that with more ozone, more hydroxyl will be present and somewhat more methane is broken down by hydroxyl. With methane being another greenhouse gas, climatic impact is somewhat abated by this. To finish the discussion about NO, some of the NO is injected into the stratosphere while aircraft spend some time above the tropopause. Also this NO produces a small increase in ozone. This extra ozone plays a negligible role in protecting us from ultraviolet light.

All the aircraft contrails put together form only a small amount of cloud compared to the area of high cloud normally present. Therefore their impact is negligible, apart perhaps from areas near very intensive air traffic. Of greater interest are the secondary effects as contrails and the particles produced my trigger the formation of further cloud cover. These high clouds, called cirrus clouds, cover a larger area and may increase the greenhouse effect. The effect may be 0 to 0.2 percent. By all accounts still a small contribution. However, it must be admitted that this is one of the more uncertain estimates of the effect of today's global fleet of aircraft.

Further we have mentioned the emissions of soot and sulfate particles. Their effect is even smaller and can be neglected. But again, also here more research is required, as the processes of condensation have not yet been researched in sufficient detail.

Main conclusions about the impact of aviation

It is fair to say that the global atmospheric environment is affected by air traffic, and that this impact is moderate at present. The impact is as expected stronger in areas of heavy air traffic, in particular in the vicinity of busy flight corridors. Further away from these regions the impact is very small. The reason for the moderate overall effect is that many of the type of pollutants produced by aircraft are already present in the upper troposphere in larger quantities, either from natural processes or from activities at the surface of the earth.

A striking feature of all the rather complex analyses of the impacts is that the uncertainties are often as large as the calculated effects themselves. This of course does not mean that one can forget about them. Rather the atmospheric system should be understood in more detail. Furthermore, it has to be taken into account that air traffic is rising rapidly, and it is desirable to do more observations and calculations. One should certainly try to avoid surprises like the one of the ozone hole, and further research is the clue to this. Below we will present some innovative approaches in this respect.

What can be done in the future and now

The present day uncertainties are caused by the complexity of the system, and this situation is aggravated by the rather small amount of observational data. Indeed there are surprisingly few measurements of NO for instance. Furthermore, better mathematical models are required. These however cannot be tested without substantially more data.

The aircraft manufacturers make major efforts to produce aircraft that meet many demands, including those of high environmental standards. Fuel efficiency has improved continuously, but certainly there are limits to this development. Furthermore, detailed research has been carried, among others by the German DLR, to study the exhaust gases in situ. Despite this, the gaps in our knowledge about the atmosphere and its chemistry are still huge. More observations are expected from new generations of instruments carried by satellites. But satellites may give a global view, but cannot measure all components present in the atmosphere that are of importance to us. Therefore we still rely very much on research aircraft flights to study the atmosphere, in different regions and at different altitudes. And this is expensive.

New initiatives for obtaining better and more atmospheric observations deserve special attention. In Japan, one JAL Boeing 747 takes small air samples during flights from Tokyo. This project has helped to better define the distribution of greenhouse gases in the remote atmosphere. Swissair has allowed the measurement of NO for one year using one of their aircraft. This has significantly increased our database for this important gas. This demonstrates how few NO data we actually did have for the remote atmosphere.

These projects use commercial passenger aircraft to help to learn more about the atmosphere. These approaches are very economical. While passengers are comfortably transported, scientific equipment quietly monitors the composition of the air. Europe has 2 more of these projects. Mozaic, supported by LUFTHANSA and Airbus Aerospace, is a major effort to get global ozone and water vapor data.

CARIBIC is a joint initiative involving LTU International Airways, the Max Planck Institute for Chemistry, the Institute for Tropospheric Research in Leipzig, and the Institute for Meteorology Karlsruhe. CARIBIC has been operational for 2 years now. The approach take by the CARIBIC-LTU project is to equip a luggage/freight container with a wide range of equipment. Two times per month, this equipment package is loaded, just like standard freight, into the cargo bay, after which the container with equipment automatically carries out many measurements. Particularly important is the fact that thanks to a special inlet system and sensitive equipment the small particles (aerosols) can be measured for the first time. The CARIBIC-LTU measurement projects is also supported by the European Commission (DG XII) as a joint European project involving Sweden and the Netherlands as well. Thanks to several institutions and nations cooperating in such a vital issue, data on particles, carbon monoxide, ozone, carbon dioxide, methane, and many more gases can now be obtained in an economical way.

The final conclusion from this all is that much work is carried out to understand the atmosphere and its chemistry better. There is no doubt about the great importance of this. It appears that under most circumstances the effects of aviation for the global atmosphere are modest, and scaled to the actual amount of fuel burned higher up in the troposphere. There are no extreme sensitivities that have been discovered. The atmosphere and its chemical systems form a complex dynamic system, of which the natural equilibria are slightly shifted. One cannot rule out surprises, as surprises are part of an incomplete understanding and information.

Finally, one can state that for mankind the sky may not be the limit, but certainly the sky has its limits. Science and technology can help us to continue to fly and have happy landings...

PART II

INTERNATIONAL AND EUROPEAN ENVIRONMENTAL POLICY: PROTECTION OF INDUSTRY OR ENVIRONMENT?

The International Perspective

Upali K. Wickrama
ICAO – International Civil Aviation Organization

Introduction

I am very pleased to be invited by the Second Hamburg Aviation Conference to provide an international perspective on aviation vs. the environment. In my presentation I will be addressing the issue of the Economic Assessment of Environmental measures.

ICAOs work on environment protection is carried out by a special committee appointed by the council, committee on Aviation Environmental Protection referred to as CAEP. The work of CAEP on any policy decision whether its related to Noise Stringency, Aircraft Engine Emissions or market bases options is guided by the following three tenets:

- Technical Feasibility
- Economic Reasonableness
- Benefits to the environment

Before I discuss the issue of Economic Assessment of environmental measures, let me look at the development of aviation traffic.

During the last 50 years, Air Transport industry has experienced rapid expansion as the world economy has grown and the technology of Air Transport has developed to its present state. As an example of this growth passengers carried on scheduled airline services increased from some 9 million passengers in 1945 to just over 1.4 billion passengers by the year 1998 at an average annual growth rate of over 10 percent. This growth rate is about 3 times the growth of world gross domestic product measured in real terms, the broadest measure of economic activity. The changes in technology and in the regulatory environment have also had great effects on the growth of air travel demand.

It is in this context that I would like to discuss the economic assessment of Aircraft Engine Emissions, ie the steps involved in the analysis of the Economic impact of a NO_x stringency proposal. This proposal describes the steps involved in the analysis of a NO_x stringency proposal directly applicable to new engine types, from December 31, 2003 and derivatives of aircraft that require certification.

For purposes of this assessment, the termAnew@ has been defined to include only derived versions of current engines and all new engines that are certified after the effective date of the proposal. Consequently, in this analysis report, We have included direct cost to manufacturers of making new or derived engines compliant. Nevertheless, an attempt has been made to estimate indirect market-driving cost to manufacturers and aircraft operators of voluntarily making existing affected in-production engines compliant.

The stringency proposal would have no direct cost impact on either the in-service or current production fleet. However, it would have a direct impact on new engine designs as well as derived versions of existing engine types. Manufacturers are actively developing technology in the near term in sufficient detail to evaluate NO_x vs. fuel burn. Within that limited window, for the engines in development, a trade-off has not been quantified. Therefore, no such costs were included in this estimate.

It should be noted there is a time-lag between the cost of implementing the proposal and the accrual of benefits of reduced NO_x emissions. Estimates of the cost of impact of this proposal do not include costs associated with the manufacturers' previous actions to meet ICAO CAEP/2 stringency levels. However, additional information provided by manufacturers of costs already incurred in anticipation of an increase in NO_x stringency beyond CAEP/2 is presented.

In order to undertake this assessment, the following information was required:

- a list of in-service and in-production engines that would not meet the proposed rule
- whether new standards would result in manufacturers modifying current products
- estimates of cost of modifying existing production engines:
  a) as a per cent of total cost of development of the engine concerned
  b) as a total per cent of the life cycle revenue of the engine concerned
- whether new standards would affect designs of engines currently in development or likely to be developed in the near future
- the scope for and cost implications of modifying designs to ensure compliance with new standards
- whether this proposal would have a fuel consumption penalty

The engine manufacturers assessed the cost impact that the current NO_x stringency proposal would have on their current and future products. In assessing the cost impact, the engine manufacturers estimated the direct cost impact for new engine types and the potential costs for voluntary compliance that could be incurred by the manufacturers and aircraft operators. This cost data is of a proprietary nature.

The following table identifies all of the aircraft engines that have Dp/Foo NO_x characteristic values that would be above the line as indicated in Figure characteristics NO_x of new and in-production engines. The engines not meeting the proposed increase in NO_x stringency, listed in the Table, have been categorized as either engines currently being manufactured and are in service, or new engine types in the design and development process. The engine manufacturers reviewed their development plans for engines that would be certified over the next five to ten years in order to identify new engine types that would be directly affected by this proposed increase in NO_x emissions stringency.

Table of Engines having Dp/Foo NO_x characteristic values above new proposed standard for NO_x emissions

New Types (Certification required)	**Production** (In-Service)
PW PW4090 (Re-cert.)	SO PS 90A
PW PW4074D	PW JT8D-217C
PW PW4077D	PW JT8D-219
PW PW4084D	PW PW2040
PWC growth engine	PW PW4090
	PW PW4164
	PW PW4168
	RR RB211-535E4
	RR RB211-535E4B
	RR RB211-524G
	RR RB211-524H

The new proposed NO_x standard, as defined earlier, would only be directly applicable to new engine types and derivatives of existing types for which the date of manufacture of the first individual production model is after December 31, 2003. If design modifications are made to an existing engine type, for which the engine type will be required to go through the certification process again, then the re-certified engine type must comply with the new proposed NO_x standard. From the Table above the engines directly affected by this new proposal would be the PW4090 (re-cert.), PW4074D, PW4077D, PW4084D, and a PWC growth engine for which there would be a direct cost impact to bring these engines into compliance. These costs will be discussed below.

It is not proposed that production aircraft engines as listed in the Table would be required to comply with the new NO_x emissions standard at any future date nor is it proposed that aircraft engines currently in production would have to comply with the new NO_x emissions standard in the future as this current NO_x stringency proposal does not state a date after which all engines being manufactured must comply. Engines out-of-production, but still in service, also would not have to comply with the new proposed NO_x emissions standard as there is no operational compliance date in the proposal.

However, market forces and potential local/regional operating restrictions might encourage the manufacturers to modify their existing products, so that they, too, comply with the proposed stringency. These product modifications could be called Avoluntary environmental enhancement@ products. It was decided to only analyze the technical and cost impact of Avoluntary environmental enhancement@ of current production engines that do not meet the proposed additional NO_x stringency. Out-of-production engine types were not considered.

Some of the production engines that are close to the new proposed limit, such as the PW2040, PW4164, and PW4168 might be made compliant through additional engine emissions testing. The engine manufacturers have provided estimates on capital costs and operational costs for Avoluntary environmental enhancement@ products for the aircraft fleets powered by the JT8D-217A, JT8D-217C, JT8D-219, PW4090, RB211-524G/H and the RB211-535E4/E4B engines. Although the JT8D-217A is not in production, thus the reason for its absence from Table 3.1, the MD-80 aircraft powered by the JT8D-217A has been included in the assessment of costs that could potentially be incurred by aircraft operators if they chose to voluntarily comply with the new proposed NO_x stringency by upgrading these engines. The reason for including these engines is that the JT8D-217A engine can easily be upgraded to the JT8D-217C model, which improves fuel burn, and operators have been steadily upgrading from the AA@ model to the AC@ model over the past few years. These costs contain product development cost estimates that would be incurred by the manufacturers and cost estimates (capital and operational) that would be incurred by the operators. These costs will be discussed in further detail in the following chapters. It has been assumed that the operational costs from the incorporation of these Avoluntary environmental enhancement@ products would be relatively small when compared to the total capital costs. For these products to be successful and voluntarily incorporated into existing aircraft engines in order to achieve emissions compliance, the operational penalties, (fuel burn and maintenance cost), would have to be nearly zero through the design and development of the product, or otherwise they will not be incorporated into the engines. Therefore, the operational penalties were assumed to be zero for this analysis. These assumptions have been reviewed with the engine manufacturers and are considered reasonable based on their past experience with other kit programs. For some engine models, the operational impact of the enhancements may result in operational savings. Where engine manufacturers have considered that operational savings are probable, then these savings have offset some of the costs ofAvoluntary environmental enhancement@.

Although there is no current production engine compliance date or operational restrictions in the NO_x stringency proposal, the issue of the impact of new emissions regulations on fleet aircraft asset values has been raised. Two studies funded, by the U.S. Federal Aviation Administration (FAA) and the other by the European Commission (EC) revealed no correlation between announcement and implementation of the CAEP/2 emissions standards and aircraft values. A review of quantitative data was inconclusive and limited data was available. During the time periods for these studies the market conditions were changing so quickly that it was not possible to determine if the promulgation of emissions rules had any value effects. The evidence from the two studies is that a large number of factors impact on such asset values.

The cost impact of the current NO_x stringency proposal over and above the normal costs of engine development to be incurred by the engine manufacturers should this proposal become the new standard has been analyzed. Additional costs that could potentially be incurred by the engine manufacturers and aircraft operators under different scenarios are also considered. The four categories of cost considered along with key assumptions and definitions are described below.

Incremental development costs for new engine types are described as those capital costs which would be incurred by the engine manufacturers to bring into compliance with the new proposed NO_x stringency, engines currently under design or development, including derived versions of current production engines. (First Category)

These costs to engine manufacturers are estimated to be approximately $ 50 million.*
Potential Avoluntary@ compliance costs incurred to manufacturers are described as the cost to the engine manufacturers incurred for the development of suitable engine enhancements or additional tests required to bring those engines currently in production into compliance with the proposed NO_x stringency. (Second Category)
These costs to engine manufacturers are estimated to be approximately $ 50 million.
Potential Avoluntary@ compliance costs incurred by aircraft operators are described as the costs that could possibly be incurred by aircraft operators to acquire and operate the engine enhancements for the current in-service fleet for which the affected engines are still in-production, in order to bring them into compliance with the proposed NO_x stringency. (Third Category)
These costs to aircraft operator are estimated to be between $0 and $ 2000 million.
Retrospective costs are described as the additional capital and development costs already incurred by the engine manufacturers to comply with an anticipated increase in NO_x stringency beyond the CAEP/2 regulation, agreed in 1991. Since CAEP/2 the on-going debate over increased NO_x stringency led to engine manufacturers developing NO_x combustion technology beyond that necessary for CAEP/2 compliance. If the engine manufacturers had not made these decisions, then the cost of re-certificating certain engine types for compliance would have led to a larger cost incurred in either category 1 or 2 above. Some engines included in the table do not meet the CAEP/2 NO_x stringency. The cost included for these engines that do not meet CAEP/2 compliance was only that associated to meet the

* All cost figures are in 1998 U.S. dollars unless otherwise specified.

extra stringency imposed by the anticipated NO_x tringency proposal, not the cost associated with meeting the earlier CAEP/2 standard. (Fourth Category)
These costs to engine manufacturers have been estimated to be approximately $ 150 million

The assessment of the cost estimates

The costs shown for the first category, represent incremental development costs that definitely would be incurred by the engine manufacturers should this proposal become the new standard. These costs are capital investments that the engine manufacturers would have to make for the development of combustor technology to make new engine types compliant with the proposed stringency.

Costs shown for the second category, represent costs that would likely be incurred by the engine manufacturers. Although current production engines are not directly affected by the current NO_x stringency proposal in that these engines do not have to be made compliant at any future date, the engine manufacturers are likely to develop engine enhancements for their current production engines which would make them compliant with the new NO_x stringency proposal. This development of the production engine emissions enhancements would only occur if the market place showed enough interest in the enhancements or if the failure to meet the new proposed stringency became a competitive disadvantage. However, provided the market was sufficiently developed, manufacturers would be expected to recover these costs from future sales.

The size of the cost impact shown in the third category would depend on the spread of charges or possible operational restrictions directed against in-service aircraft not compliant with the EPG proposal. These charges or restrictions could be enacted at the local, regional or global level. The potential cost, therefore, could vary from almost zero in the case of scattered local regulations to as high as $2,000 million dollars in the case of global restrictions however this represents aAworst case@ scenario. Such a scenario could only be regarded as plausible in the event of operational restrictions introduced at the global level or widespread local restrictions introduced by most states and airports. In the absence of such measures, the view of FESG is that any voluntary environmental engine upgrades by the operators to make engines compliant with the proposal will be very limited. There is no available evidence that operators have upgraded engines in response to the CAEP/2 standard. In the event that emissions charges are introduced by some airports, airlines may decide to incur the charge rather than upgrade the engines. In such cases the economic penalty would be less than the cost of upgrading the engines. Alternatively, aircraft operators may decide to re-schedule their fleet to operate into those airports with compliant aircraft if operational flexibility exists within the fleet.

Costs shown in the fourth category have not been included in the economic impact assessment of the current NO_x stringency proposal presented above. However, they have been included in this report to remind CAEP of the incremental capital expenditures the industry has incurred and has not fully recovered in order to reduce NO_x emissions for certain engine models well below that which was required by the CAEP/2 standard, in anticipation of further increase in stringency. The costs included here are the part of the technology development costs that was associated with the anticipated additional increase in NO_x stringency below the CAEP/2 standard. It should be borne in mind, however, that the modification of an engine to meet more stringent emission rules will seldom be pursued purely to meet a tighter limit, except perhaps for those engines for which the non-compliance is marginal. It could result in additional engine sales, extend the spares sales stream and maintain the resale value of an aircraft. The costs are therefore not necessarily lost, nor total debits from a manufacturers profit, in the event such investment potential is realized.

Benefits Analysis

Benefits from new stringency are difficult to assess, for various reasons. Since there is as yet no consensus on a monetary value that might be attached to atonne of NO_x estimates are normally calculated in terms of tonnes of NO_x saved. The development of better NO_x technology whilst optimizing fuel consumption is a difficult technical balance which the engine manufacturing industry have to address in the context of overall performance of their products. This has influenced the scale of proposed stringency measure which represents a technology improvement for the engine industry. The costs of the proposal, if implemented, are estimated to be small, and the benefits are expected to be similarly limited. The costs of complying with tighter standards tend to take place over the short term, as engines are re-designed or retrofitted with appropriate modifications. On the other hand, any benefits in terms of reduced NO_x burden will accrue over the long term, as compliant aircraft slowly enter the fleet, and the non-compliant aircraft retire.

For this proposal, there is a direct relationship between costs and benefits, in that there would be real NO_x reductions from each and every engine designed or modified to make it compliant. These benefits would materialize for all new engines requiring redesign under the proposal. Since the proposal does not contain a production cut-off date, there would be no requirement to modify noncompliant in-production engines. However, it is thought likely that these would be modified for competitive market reasons, and for each such engine produced there would be a direct benefit in terms of reduced NO_x emissions. Similarly, for any in-service engine modified for compliance, there will be a direct NO_x benefit, although it is considered unlikely that there will be large numbers of affected inservice engines modified.

Unfortunately, detailed quantitative global estimates of the benefits resulting from the proposal could not be produced in the time available. The global benefits estimated by looking at the effect of the proposed measure on fleet NO_x performance starting from a known base. Such analysis would have required a judgement on future market shares for the engine manufacturers, as well as estimates of airframe/engine combinations for future forecast horizons. The development of an agreed scenario encompassing such issues required more time.

However, at the present time, large majority of modern engine types in production and entering service are known to be compliant with the proposed rule, in some cases by a significant margin. Clearly, since compliance is generally an established fact, the benefits of the proposal in terms of reducing the global emissions burden will be marginal. There will be modest costs, and modest benefits, attached to redesigning the new types affected, or modifying any production or in-service engines.

But this view of the benefit potential misses the important point that this situation is the culmination of continuous pressure for change in the NO_x standard over the past decade, of which the proposal is the latest step. The NO_x stringency recommended by CAEP/2 will have influenced thefleet=s emission trend, as would the subsequent (unsuccessful) stringency proposals brought to CAEP/3. The industry's successful response to that pressure over the period has resulted in the compliant situation which now exists.

Therefore, even in the absence of an agreed quantification of global NO_x saving, the most important benefit from adoption of the EPG proposal will be the setting of a minimum NO_x technology limit at a level which is demonstrably achievable today, ensuring that future engines will not compromise past technology gains. In effect, this proposal brings the standard into line with current technology.

Voluntary compliance will also produce emissions benefits. Under the third category, it is noted that there was a potential category of cost incurred by operators who might voluntarily choose to comply with the proposed stringency measure. These operators would be those using aircraft fitted with engines currently in production, but not compliant with the proposed new measure. In the event that these operators upgrade such engines on a truly voluntary basis, it must be noted that the benefits to these operators must equal or exceed the costs of compliance. In addition, there would be the benefits associated with the reduction in global NO_x burden. Therefore, in the case of the **voluntary** costs, it can be deduced that benefits exceed costs.

In conclusion the assessment of the NO_x stringency proposal indicates that the direct costs would be minimal, and the benefits would be modest. However, the proposal appears to represent an economically reasonable step to set the technology standard at what is achievable today.

Towards an efficient market-based environmental aviation policy – A view from the Hamburg Region

Hans-Martin Niemeier
State Ministry of Commerce of the Free and Hanseatic City of Hamburg

Introduction

Hamburg Airport (HAM) is of major importance for the economic development of the Hamburg region with its 3 million citizens. The Airport has performed an enabling function and become a regional cluster for aviation-related business such as hotels, freight services and in-flight catering. One prime example is the establishment of the maintenance and overhaul center of Lufthansa German Airline in 1954 on the airport premises depending on the use of the airport's runway system. This Lufthansa subsidiary employs more than three times the number of personnel of the HAM Corporation. Various suppliers of airport services have opened up their business off-airport.

A service and logistic center such as Hamburg relies on a high quality of complementary air services. A survey of 400 companies using HAM confirmed that good access to international air services is an indispensable condition to attract business in the emergence of a global economy. At the European level, the business community demands more frequent and better air services to and from various locations throughout Europe. If Hamburg is to not lose ground to competitive airports or regions, it is essential to operate daily flights to and from as many European cities as possible. The frequency and accessibility of air transport services has improved thanks to the liberalization of the European Community's internal air transport market. Barcelona, Rome, Basle, Strasbourg, Eindhoven and Manchester were among major European cities, which were not served directly from Hamburg before the liberalization in 1985. Ten years later, when the second stage of liberalization came into effect, all these cities were served with return flights on working days, allowing at least a five-hour stopover for business affairs.

Regulatory liberalization in domestic and international markets and the tremendous economic growth experienced in North America and Europe has led to an increase in environmental impacts of aviation as well. Externalities of air pollution and noise impacts have increased with the growing demand for passenger and freight. In general, external diseconomies occur when the economic agent responsible for the activity does not face the full costs that result from the activity. In a free-market economy, economic agents typically only maximize their own utility or profit functions without regard for external costs and benefits. These externalities will not be reflected in the market price of the commodity or service. In an aviation context, the industry may make air travel inexpensive, but nevertheless be polluting the atmosphere and generating significant noise in the process to the detriment of non-air travel consumers. Unless the full cost of this pollution is reflected in the price of air travel, people will buy more than they would chose if the full costs of supplying aviation services were reflected in the price charged. This market failure causes an inefficient use of resources and has negative consequences for welfare. In order to improve welfare, economic policy must not only liberalize the aviation industry but must also internalize the negative external effects of aviation (cf. MISHAN, 1984).

Noise pollution makes up most of the local negative external effects of aviation at HAM and its vicinity, while other polluters, such as carbon dioxide (NO_x) emission, are of still of lesser importance in spite of an expected NO_x increase. Consequently, noise issues of Hamburg's inner city airport are at the center of public discourse of the City State of Hamburg and its neighbor, Schleswig-Holstein. The overall objectives of a joint aviation policy of these two northern states of the German Federation are to overcome these inefficiencies by a market-based environmental aviation policy as outlined in the Guidelines for a Northern German Aviation Policy (LEITLINIEN, 1995)[1].

Obviously, local and global externalities are closely linked. Modern aircraft and their engines are generally less noisy and have lower emission levels which is less damaging to the global atmosphere. If the use of modern aircraft can be fostered by the price policy of airport charges, then regional aviation policy contributes to addressing problems of the global atmosphere[2]. Complementary, airports and their regions would benefit from efficient international measures fostering the use of modern aircraft and, thereby, reducing global emission levels. Such measures would reduce noise levels at airports and would reduce the magnitude of local external effects.

---

[1] The theoretical foundation of the guidelines as well as of the market based environmental policy in this paper is standard microeconomics. It is critically discussed in NIEMEIER (1998)

[2] If all airports try to internalize local external effects, high emission aircraft would be shifted to those airports where social marginal costs are low. Such a shift would have a positive effect on economic welfare. Furthermore, I think that it would also have a positive effect on the aggregate level of global emissions, as operation costs of aircraft with low emissions would be lowered relatively to aircraft with high emissions.

Economic, environmental and political factors are intertwined in a market-based aviation policy, but there are boundaries of responsibilities. The objective of regional economic policy is to internalize the negative external effects of local aviation. Complementary, national economic policy should address the global external effects in collaboration with international regulatory bodies, such as the European Union and ICAO. As Tinbergen has already pointed out in the fifties two goals of economic policy, namely internalizing local and global externalities, cannot be reached with one instrument. It is necessary to use efficient measurers at both levels.

The objective of this paper is to outline how Hamburg's aviation policy tries to find efficient solutions to environmental problems. From my point of view, the solutions could serve as a role model for other airports. Initially, I would like to address the impact of Hamburg Airport on its regional economy, followed by the interdependence of passenger growth and noise pollution. These two sections deal with the positive and negative effects of aviation and show the 'dilemma', which the aviation policy of Hamburg faces. Thereafter, I am going to outline the goals of Hamburg's Aviation Policy and discuss the related implementation strategy which tries to balance the negative and positive effects of aviation. I will introduce the noise budget for HAM, which is the basis of our market-based environmentally sound aviation policy. Finally, I will close with the issue of pricing instruments at the European and international level as local environmental measures might be effective if global externalities are efficiently internalized and more market-based options for local externalities are prescribed by international law.

Regional Economic Impact of Hamburg Airport

Evaluating the importance of an airport to a regional economy is difficult. There are demand side and supply side effects of various natures. The direct economic contribution is generally measured using economic impact analysis which provides labor and income indicators.[3] I confine myself to referring only to the employment results (cf. ARING, cf. page 98).

The direct employment, defined as the number of jobs at the airport, amounts to almost 12.000 jobs. In 1995, the airport corporation itself employed 1.500 personnel and airlines (operating at HAM) 1.200 personnel, while the Lufthansa Aircraft Maintenance Center accounted for 5.500 jobs and various other tenants for 3.800 jobs. In contrast, the indirect and induced employment effects are not

[3] The Hamburg Ministry of Economic Affairs has contracted a study on the economic impact of HAM on its region which was conducted by empirica Ltd., Bonn, Germany in 1994 (ARING ET AL., 1995).

known unless a regional airport impact assessment is undertaken. The indirect impact is the labor content resulting from the supply of goods and services for the companies at the airport. The inputs from the region are estimated with the help of a regional input/output model. Direct and indirect employment generate income, which generates additional demand for goods and services and induces employment in the region and in other sections of the economy.[4] The indirect and induced employment effects of HAM involve about 20.500 jobs worldwide and 15.600 jobs in Hamburg's Metropolitan area according to the empirical study. The total employment impact is the sum of direct, indirect and induced employment. It consists of 32.500 jobs worldwide and 27.600 for the Hamburg region.

In the public debate about aviation and environment, economic impact studies play an important role given the high level of unemployment in Europe. Gazzard (Friends of the Earth) in his presentation "The Ecological Perspective" pointed out that in the controversial debate about airports and their environmental problems, it is not unusual for airports to refer to their regional employment impact.[5] Airports argue that unless you stop demanding noise and pollution reductions at the airport, you may loose your job and endanger those of your neighbours.

Such logic stems from a misleading interpretation of the results of economic impact studies. Theoretically, the positive employment effects derived by input-output analysis are only lost if the whole airport would be closed and would not be substituted by other economic activities[6]. Environmental policy does not demand such losses but instead aims at substituting noisy and polluting air services by modern 'environmental friendly' air transport[7]. Also, such claims do

[4] The induced effect is determined by the Keynesian multiplier analysis.

[5] Of course, it is tempting to use regional economic impact assessment to boost the employment figures by applying high regional multipliers (cf. MUELLER, 1998, ARING ET AL., 1995, page 104). Some airports' claim of their total employment seems exaggerated simply because the region was defined very large or a year with a high investment level of the airport was selected. Although the magnitude of indirect and induced impacts is often doubtful, the denial of any indirect and induced employment effects is certainly not credible either. The Hamburg Airport Study was contracted with the premise to avoid an overrating of the employment effects. Thus, empirica ltd. applied conservative assumptions to its model. In Hamburg we have so far abstained with some minor exceptions (cf. FLUGHAFEN HAMBURG, 1996) from the temptations to overstate the employment effects of HAM.

[6] On the limitations of input-output analysis cf. PFAEHLER and GABRIEL, 1999; AMSTRONG and TAYLOR, 1993; TSE and TRETHEWAY, 1995.

[7] Of course, one also has to look at the price attached to environmentally friendly transportation, but I doubt that efficient measurers will cause negative employment effects as the propensity to consume and to invest seems to be not negatively effected.

no rely on sound empirical facts. As one could observe from the historicaltrend, economic growth and environmental protection can go hand in hand. Operations at HAM today result in less noise pollution and lower emission levels than previously, but create more jobs than 10 or 15 years ago. This outcome is a consequence of both sound environmental policy and a burgeoning economy. Aviation policy is committed to economic growth but makes also a commitment for some environmental protection by creating incentives for the substitution of noisy and polluting aircraft with those aircraft that are more 'environmentally friendly'.

The Ministry of Economic Affairs developed an economic aviation policy that aims at ensuring planning and investment security for both the airport and the adjacent business community. Aviation policy and urban planning share the goals of encouraging the establishment of a cluster of aviation-related business that conforms to environmentally sound economic development.

Passenger growth and noise emissions at Hamburg Airport

The economic benefits of better air service among major European cities did go along with a substantial increase of noise at the beginning of this process.

From 1985 to 1990, the total number of passengers at HAM grew from 4.8 to 6.9 million - an increase of 44 per cent; commercial aircraft operations rose from 73.000 to 112.000 - an increase of 55 per cent. At the same time, the overall noise levels (Leq-3 by 2-3 dB) rose sharply too, primarily because the fleet mix of noisy aircraft remained nearly unchanged. In the nineties this trend was reversed: growth in passenger traffic did not necessarily lead to increasing of aircraft noise; on the contrary, noise level could be reduced (compare exhibit 2). Larger aircraft and higher flight frequency have accommodated passenger growth. The larger aircraft in particular tend to be Stage III and are quieter and have lower emission levels.

Exhibit 2: Passenger growth and reduction of noise level at HAM

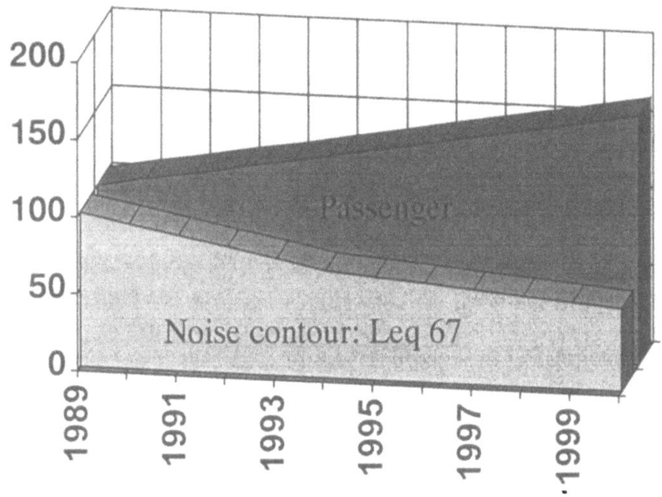

According to earlier passenger traffic forecasts for HAM covering the period from 1990 to 2000, the number of passengers was expected to increase roughly[8] by 60 per cent from 6.8 million to 10 million passengers (cf. ISERMANN ET AL., 1992). During the same period, the area exposed to a given constant average aircraft noise level, measured in Leq, was predicted to shrink by half. The actual development thus far confirms the forecast trend. Aircraft operations were expected to grow by 20 per cent as a result of utilizing larger aircraft with higher seat capacity, along with improved load factors. Meanwhile, slow and noisy aircraft have been gradually replaced by less noisy technology. At HAM, the share of Chapter 2 aircraft in total jet movements is to decrease from 60 % to 5 % in the period 1990 to 2000. These developments contributed to reversing the trend of the eighties: passenger traffic growth does not necessarily result in higher noise levels. Noise levels have been reduced to the levels of the eighties when the airport handled less than half the number of passengers.

These positive developments of the nineties do neither imply that noise pollution is marginal at HAM nor that the noise problem will vanish automatically in the future. However, there seems to be less tolerance of people versus noise. A crucial question for the acceptance of HAM by the local community is whether a strong demand for air service will eventually increase noise levels again? To be

[8] For a detailed discussion compare NIEMEIER, 1998.

more specific: According to a new forecast commissioned by Hamburg's Ministry of Commerce to MKmetric Ltd. (MANDEL, 1997), HAM will handle 13,2 Million passengers and 172.000 aircraft operations by the year 2010 (see exhibit 3 below)[9]. Can the environmentally favorable trend between growing passengers traffic and reduced noise levels be maintained when passenger numbers are expected to increase from currently 9.2 million by about 4 million within in ten years? Is it realistic to hope that a constant noise level could be maintained when about 50.000 more aircraft movements are expected during the next decade? Or are we confronted with the fact that more noise is an inevitable sacrifice in order to accommodate future passenger demand?

Exhibit 3 Passenger growth at HAM

These questions were debated over the last years when the airport was going

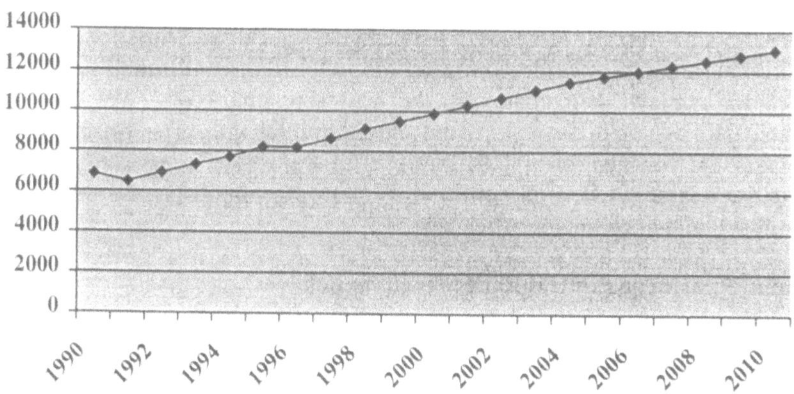

through a mandatory planning process to obtain permission from the local community to increase apron and terminal capacities. The next two sections will outline what strategy was adopted by the regional government to address these challenges.

2. Goals and Strategies of Hamburg's Aviation policy

Having stressed the economic impact of Hamburg airport on its region, it is understood that the enhancement of airport performance is a major goal of the city-state of Hamburg (see exhibit 4). Improved air services are vital for the continuation of the competitiveness of the service-based economy of Hamburg. To enhance airport performance it is necessary to use the existing airside capacity more efficiently and to overcome bottlenecks like the cramped apron.

---

[9] A multi-modal system approach was used for this forecast (cf. MANDEL 1997 and 1999).

Exhibit 4: Goals of Hamburg's Aviation policy

- Enhance airport performance
- Environmental protection
  - Internalization of external effects
  - Contribution to CO_2 reduction of 10 per cent until 2005 based on CO_2 emission of total German traffic in 1987
    - Efficient displacement of loud aircraft
    - Efficient use of technology for reduction of noise and emissions

The goals of environmental protection and enhancement of airport performance should be treated in a coherent way. External effects should be internalized to increase welfare.[10] Furthermore, aviation should contribute to a 10 percent CO_2 reduction over the 1987 level of all German transport modes by the year 2005. In order to fulfil the environmental protection targets, loud aircraft need to be replaced by new aircraft technology, reducing both noise and emissions. A fleet substitution program should be designed not only in a non-discriminatory way but should achieve their goals with the least economic costs. We have not assessed the costs and benefits of our environmental measures quantitatively (compare the contribution of GILLEN) but instead based our strategy on a qualitative reasoning. Let me outline this strategy (exhibit 5) and then come back to the point of economic efficiency.

Exhibit 5: Strategies of Hamburg Aviation policy

MARKET BASED		INFRASTRUTURAL/ ORGANISATIONAL	
Tax	Charges	Extension of apron and terminal	Noise protection program
Certificates	Noise budget	Environmental management	Land use planning

The strategy of Hamburg's aviation policy consists of two elements: Market-based instruments combined with infra-structural and organizational measures. Due to German federal as well as European legislation, the regional government cannot implement two out of four market-based options, namely green taxes and

[10] External effects should be internalized because it increases the economic welfare as every economic student learns in its first semester (strangely there is debate among transport economist in Germany compare ROTHENGATTER, 1994). Furthermore, this goal is not only in the altruistic interest of welfare maximizing public body but also in the more self-interested body of a city-state. Social costs are partially borne by the city in form of less attractive urban districts, which promotes sub-urbanization with negative effects on the financial budget of the city-state

tradable certificates. Therefore, Hamburg aviation policy relies on noise charges and a noise budget. Noise charges, distinguish between noisy and less noisy aircraft, whereas a noise budgets[11] limits aggregate noise emissions for the airport. These measures are combined with an expansion of the apron and terminal building, with active environmental management, noise protection and land use programs (see the contribution of SCHMIDT).

The strategy does not include administrative measures, such as banning selective loud aircraft, stricter general night-flight restrictions and the prohibition of charter traffic and general aviation. It relies instead on market-based instruments that try to influence emissions through the price mechanism. Such a strategy has four advantages (see LEITLINIEN, 1995):

- It does not interfere with the principal business decisions of airlines and consumers, which can adapt flexibly by way of choice. No options are ruled out, only relative prices are changed.
- It minimizes the cost of the measure because those airlines, which can afford to invest in noise reduction technologies with the relatively lowest costs, contribute most to the noise reduction.
- It gives an incentive for those airlines which have invested in less noisy aircraft and have achieved high load factors.

It encourages technological progress because it sets signals to develop aircraft with progressively less emissions.

Noise budget for Hamburg Airport

It is very popular to demand movement restrictions in order to decrease noise. In Hamburg for instance, the Green Party demanded a freeze on movements at current level in its election program, but there are good reasons why such a measure might be inefficient compared to a noise budget (see exhibit 5).[12]

Exhibit 5: Noise budget versus movement restrictions

	NOISE BUDGET	MOVEMENT RESTRICTIONS
ADVANTAGE	▪ Incentives for ➢ less noisy aircraft ➢ Higher load factors	▪ Incentives for ➢ larger aircraft ➢ higher load factors

[11] Of course, it would be optimal to extend the noise to a general emission budget divide the budget into pieces and design a certificate auction in combination with a slot auction. The LEITLINIEN recommends this as a long-term option as it needs a new European air law framework (cf. also KUNZ 1999, WOLF, 1999 and BUTTON, 1999).

[12] In this comparison I assume that efficient allocation mechanisms such as differentiated charges are used to enforce the restrictions. The advantage of a noise budget are lost for instance if restrictions such as stricter night curfews or a banning of chapter 2 aircraft are used instead of higher landing charges for loud aircraft.

	▪ planning security	
DISADVANTAGE	▪ minimal operational disadvantages	▪ no incentives for less noisy aircraft ▪ increase of noise ▪ unnecessary economic losses ▪ reduce service quality to air travelers

One has to admit that movement restrictions have also advantages. Airlines would have a strong incentive to use larger aircraft and utilize the seats effectively, but there are sever disadvantages. There are no incentives to use modern aircraft, which are generally less noisy than old or existing aircraft. Nor is there any incentive to utilize noise reduction potential by substituting for e. g. chapter 2 aircraft and hushkits with modern aircraft. The takeoff footprint of a chapter 2 aircraft is seven times larger than a modern chapter 3 aircraft. The differences for landing are not as extreme, nevertheless the footprint is still more than double the great. However, a strategy of restricting operations does not take account of the noise characteristics of those aircraft that do use the airport. In a situation with high excess demand and high noise differences of aircraft fleet movement restrictions might even lead to an increase of overall noise levels. Noise restrictions might not even reach its aim and therefore should not be pursued any further.

A noise budget has a number of advantages. First there are incentives to use larger aircraft combined with higher load factors and this may lead to a reduction in the number of operations.[13] Secondly, it has the advantage of promoting the use of less noisy aircraft. This advantage is decisive and on the whole it seems evident that such a measure is beneficial. The operational disadvantages include monitoring costs and the cost to the airlines of utilizing the less noisy aircraft in their fleet for serving HAM. The benefits to the airport users and the neighboring communities must be mitigated by the losses to those near those airports where the noisy aircraft are shifted. These shifts tend to lead to a more efficient use of airports in Northern Germany, as these airports are located in rural areas with comparatively lower noise standards. The airport and the airlines have a lower risk-planning horizon since they know about the future environmental constraints and the aims to be reached in order to gain public acceptance. Airport neighbors in general and those who might decide to live and invest in property there know what upper noise level they have to bear in the future.

[13] This is not necessarily the case since a carrier could replace its stage II aircraft with all Stage III and reduce noise emissions sufficiently to permit an increase in flight frequency.

Although a noise budget has such clear advantages compared to movement restrictions it is not very popular. At some airports noise budgets are not effective because they are misleadingly constructed perhaps due to the interference of vested interest groups. Another reason is that the noise budget might be simply too high. Bremen Airport has a noise budget, but it is so high it would be unreasonable to think that the growth of air transport service will ever reach such a level (LEITLINIEN, 1995). The noise budget at London airports allows more movements than the movement restrictions (ISERMANN ET AL., 1997). Such noise budgets would not make any sense in Hamburg simply because they would not gain public support. Yet another reason for the unpopularity of a noise budget might be that too much is expected from a noise budget. A noise budget does and cannot solve all noise problems at an airport. Other measurers such as passive noise protection programs must accompany it. A noise budget should be a part of a consistent and efficient multi-dimensional noise protection strategy[14].

What can a noise budget achieve? Noise is sound that annoys and disrupts. It has an objective and a subjective dimension. Noise at an airport is determined by several factors: Of course acoustic factors play an important role, but so to do psychological, socio-demograhic and economic factors. Therefore, not every neighbor will evaluate the effects of a noise budget in the same way. Noise is and more important will be valued differently over time. Furthermore, a noise budget is only designed to fix the overall sound energy at the airport. It seems to be nearly impossible or at least not very practical to regulate the division of the overall energy. Compare two different scenarios displayed in exhibit 6.

Exhibit 6: Comparison of contours

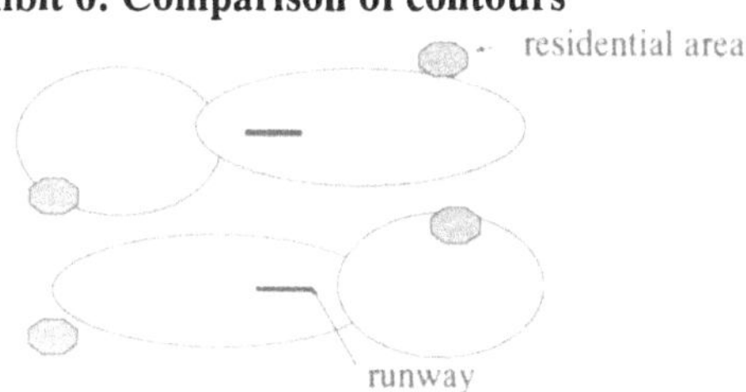

[14] The noise budget of HAM does not intend to solve the problem of noise at night. The noise budget is based on the Leq. Such an average of different noise levels does not rule out few but extremely loud aircraft movements, which might lead to severe sleep disturbances. One option is to establish a separate noise budget based on the number of loud events at specific times. Another option is to decrease the noise by enforcing the differentiation of noise charges at night times. The airport and the City-state of Hamburg have so far selected the last option.

In both cases the overall sound energy is of the same amount but has different contours. Such a change might occur when noise reductions over time are unequally distributed between landing and take-off. In case A, the residential area right above is outside the contour while the area to the left below is crossed by the contour. In case B, the situation is exactly reversed. In order to decide which case is better, one has to know exactly where more people live. This is not so difficult given the very detailed statistics of inhabitants in Hamburg. It gets difficult if you want to take account of the noise shelter of houses, of the age and health structure, of the living habits of people (for instance how long people usually stay outside and inside of their homes). Still the problems how people evaluate noise and how the marginal utilities of different people should be compared will arise.

For these reasons it seems better not to regulate the division of noise but only to set the overall sound energy. Therefore the noise budget of HAM has no fixed contour (cf. exhibit 7).

Exhibit 7: Principal components of the noise budget
▪ Based on six months of 1997 with the highest noise levels
▪ No fixed contour
▪ Leq-3 of 62 dB and manual with the most recent aircraft types
▪ Annual control
▪ Active noise management of airport through noise charges
▪ Co-ordinated capacity is reduced if noise budget is exceeded continuously
▪ Should be developed in consultation with community and airline representatives

From the year 1999 onwards, the noise level at HAM is not allowed to exceed the level of 1997.[15] This is the general idea of the noise budget. To be more precise, every year the noise will be monitored against the noise budget which is

[15] In Hamburg the level of the noise budget is determined by a political decision. From the point of welfare economics such a decision should be determined by a cost and benefit study (cf. the contribution of GILLEN). However such a study was not undertaken in Hamburg given the limited amount of time and resources. The standard price approach of environmental policy was adopted (cf. BAUMOL, OATES, 1988). The decision was based on a qualitative balance of the advantages and disadvantages. The noise budget should allow the airport to utilize its growth potential, but without an increase of noise (see below).

based on the six months in 1997 with the highest noise levels. For the base year 1997 and for the future years the noise at a level of 62 dB Leq is calculated according to the standardized official manual for noise calculation in Germany.[16]

An increase of noise exceeding this budget is supposed to be prevented by the active noise management of Hamburg airport through noise charges. If the noise budget is continuously exceeded, the coordinated capacity will be reduced. As the noise cannot be influenced directly, like in the case with a budget combined with tradable permits, an overshooting for a limited time of 2 years must be allowed. The period of 2 years might seem too long for the effectiveness of the noise budget, but such fears are not well founded. Neither the airport management nor the airlines would take a risky strategy of overshooting the budget. The possible loss of scarce slots of high value and the loss of public acceptance would prevent the airport and the airlines adopting such a strategy.

The active noise management relies on the effectiveness of price differentiation, which have been very often criticized as not effective by the aviation industry (compare e.g. STARKIE, 1985 on the discussion of the effects of price differentials among London airports). Indeed, quite often price differentiation is obviously not strong enough to induce the desirable corrective action. This has been the case for the HAM at the beginning of the nineties. According to the study of DLR for the Northern German States (BEYHOFF ET AL., 1992), the differentiation between chapter 2 and chapter 3 aircraft was only effective for German domestic traffic. Therefore, the Guidelines put the reform of landing charges on the agenda (LEITLINIEN, 1995). Since then, the differentiation has been strengthened between chapter 2, chapter 3 and bonus aircraft at HAM. Also the differentiation for general aviation has been split and became stricter. Today the surcharge of 400 % for chapter 2 aircraft (see exhibit 8) seems to be very effective as it gives a bonus aircraft an competitive advantage of roughly 45 DM per passenger.[17] On the agenda is still a structural reform of noise charges. Charges should thinly differentiate between noisy and less noisy aircraft. Every decibel counts. The differentiation should be based on measured (average) noise at the airport. The differentiation between day and night times should also

[16] In order to ensure an objective calculation the manual is updated with the most recent aircraft types. The calculation is done by the Deutsches Zentrum für Luft- und Raumfahrt (DLR), an independent German research Institute.

[17] Since 1992 HAM has increased the surcharge for chapter aircraft from 100% to 400%. Assuming a load factor of 65 % a typical chapter 2 aircraft like the Boeing 737-200 would have to pay 45 DM per passenger more than a bonus aircraft such as the Boeing 737-500.

become more effective. In short, there seems to be no principal obstacle to turn charges into an efficient instrument (MANDEL, 1997).

Exhibit 8 Aircraft landing charges at HAM[18]

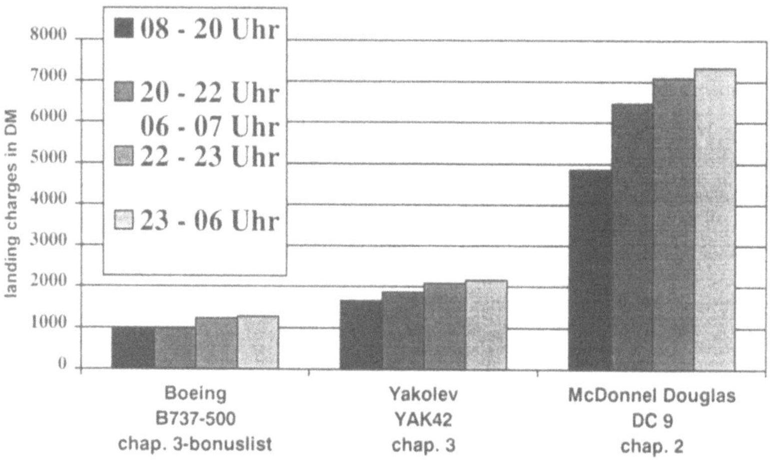

An efficient market based instrument is needed for the future development of HAM. The noise budget rules out scenarios with increasing noise emissions, which were developed by the airport in the planning process for the permission to extend the apron system (FLUGHAFEN, 1996). Such scenarios were based on self-contradictory forecasts (WIRTSCHAFTSBEHOERDE, 1998) with unrealistic high noise emissions. They might have contributed to the decisions of the red-green coalition to limit the amount of noise at a current level. The limit is set on such a level that it is a fair compromise between environmental and economic goals[19]. The noise is fixed at a historic low level (see exhibit 9). To my knowledge it is compared with other German and European airports the toughest noise budget. On the other hand it allows the airport to develop its long-term potential of 13.2 million passengers. Growth is allowed as long as additional passengers do not cause an increase of noise. An efficient charging policy which benefits larger aircraft with high load factors and modern less noisy aircraft will secure the realization of the development potential. The aviation policy of the city-state of Hamburg gives investment security for the

[18] In order to compare charges for the aircraft a MTOW of 54 t is assumed. I thank Klaus Koehler for this comparison.

[19] In order to gain public acceptance for this compromise the concept of the noise budget was discussed intensively with representatives of airlines and of neighboring communities.

airport to enhance the development as a regional aviation cluster. It is evident that this is important for the ongoing partially privatization of HAM.

Exhibit 9 Contour of Leq(3)[20]

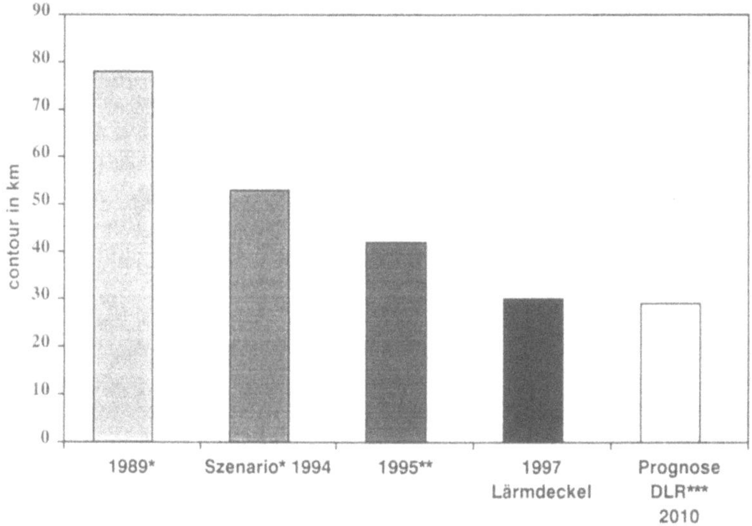

* Lärmphysikalisches Gutachten 1992
**Planfeststellungsantrag
Vorfelderweiterung 1996
***Prognose Mkmetric/DLR 1996

2. Pricing Instruments at European Level

From a regional standpoint the current environmental policy at the federal and European level is ineffective, inefficient and an obstacle for optimal solutions of local environmental problems. Let me explain this sharp criticism (for a more diplomatic, but in substance equally critical view see LEITLINIEN, 1995).

Firstly, there are no market based environmental instruments in effect in the EU aviation market. The EU policy relies wholly on administrative or regulatory instruments, such as Directive 92/14/EEC limiting the operation of Chapter 2 aircraft at Community airports by April 2002. These regulatory instruments have well known deficiencies, namely no incentives to limit noise and emissions

[20] The contours of 1989, 1994 and 1995 would be smaller if the latest manual for noise calculation had been used. Such calculations are not available, but would not alter the trend of decreasing contours. I thank Klaus Koehler for this comparison.

below the chapter 2 standard, but strong incentives to meet these standards exactly by hush-kitted aircraft. The large potential of noise reduction within chapter 3, which the noise reduction policies of airports should have used in the 90's and which airports must use in the future, is not utilized. The current EU proposal to limit hush-kitted aircraft faces the problem that it has created by its own policy. Airlines that have invested in hush-kits to meet EU standards feel now as if they are being punished. Furthermore, as modern aircraft would have replaced most chapter 2 aircraft anyway, the effectiveness of the directive seems to be rather low. It has simply had the minor effect of advancing displacements that would have occurred anyway. But does it make any sense to set these limitations at all? The impact of these directives on the future noise level at most European airports such as HAM, seems to be nearly intangible. At the same time, the operational opportunity for airlines with chapter 2 aircraft is totally restricted although there are airports where chapter 2 aircraft would not disturb the public.[21]

Secondly, the EU policy does not abolish hidden subsidies and does not tax fuel of commercial aviation. There are no instruments currently in place that try to internalize global externalities (compare EUROPEAN COMMISSION, 1998).

Thirdly, EU law hinders optimal[22] solutions to environmental problems at the local level. It does not allow the establishment of a noise and emission budget with a well-designed slot and certificate auction. Such instruments would increase the effectiveness of local environmental protection. These instruments could even improve the efficiency of the whole EU aviation market if European airports use such market-based options instead of administrative, but inefficient instruments.

21 The policy of the EU should not try to establish equal noise standard at all European airports because externalities among European airports differ. Externalities at inner city airports are obviously higher than at airports in rural areas. Therefore the EU should not try to solve such problems with general measurers such as the outfacing of chapter 2 aircraft. The European aviation policy should aim at global externalities and use efficient instruments.

22 Since the beginnings of the seventies Bonus has argued that standards should be combined with permits because such markets are environmentally and economically efficient (cf. e.g. BONUS, 1984). Such claims seem to be slightly overstated (cf. BAUMOL/ OATES, 1988). From my point of view, whether differentiated charges or certificates are more effective instruments depends on how the rules of permit markets are designed and regulated. Since first and second best solutions are nearly out of reach I recommend that permits and slots are traded and the rules of such trade be improved.

For these very reasons, the EU leaves the burden of environmental strategies to the regional airports and their municipalities. The guidelines of Northern German aviation policy therefore demand that efficient pricing instruments be adopted at the European level. A fuel tax and/or an emission tax should be adopted as well as slot and certificate trading. Of course, these instruments should be designed in a coherent way involving other airports as well as other modes of transport in order to have a fair and efficient competition (see exhibit 10).

Exhibit 10: Pricing Instruments at European Level		
Guidelines for Northern German Aviation Policy		
Market based options at European level		
FUEL TAX	SLOT/CERTIFICATE TRADING	EMISSIONS TAX

3. Concluding Remarks

Global competition, privatization and liberalization are the buzzwords for corporate strategies and regulatory frameworks of the nineties. As far as environmental problems of aviation are concerned, they reflect an increasing pressure to find economic efficient solutions. Only if these efficiencies are gained it seems possible to win public acceptance and be competitive. Privatization together with effective price cap regulation (cf. KUNZ 1999) puts a price tag on airports and those airports that adopt efficient instruments will win. I have argued that the neoclassical concept of internalization is a useful framework to find such optimal solutions. The noise budget together with efficient charges is a good instrument to internalize external effects. It is far from being perfect, but from my point of view internalization is a tantonement process in which one under and over shoots the target. The efficiency of European airports could be increased if such concepts are adopted by a growing number of airports. Charges would then reflect the marginal social costs and airlines could choose freely which airport with which type of loud or less noisy aircraft should be served. As optimal pricing of peak-hour traffic is an instrument which would lead to an efficient use of "physical" airport capacity, so analogously marginal social cost pricing would lead to an optimal use of "environmental" capacity. Airlines and passengers would win, as comparatively high costs of restrictions would be avoided. Such a price regime for airports would internalize local externalities, but not global social costs. Therefore, it is necessary to use market-based options, such as emission taxes and certificate trading at a European and worldwide level.

The German environmental policy of the seventies and eighties relied on regulatory instruments. Holger Bonus, one of the most outspoken economists of a market based protection of the environment, once said with resignation that not economists, but “lawyers and engineers invented the instruments” (BONUS, 1992). In Hamburg we reformed our environmental policy along a market-based approach for the very reason that only an efficient airport with sound environmental management will increase the economic welfare of the City-state of Hamburg. Reading the guidelines of the federal and EU transport policy, one gets the impression that finally the dawn of the economists century begins after a long period of policies based on regulatory instruments. There is widely shared consensus about policy goals: sustainability and internalization of externalities. But turning to the concrete measures one feels disappointed. Up to now no efficient means for attaining these aims are implemented. Therefore, I fear that the century of the economist has not yet started and I get the impression that many of my fellow economists are just being hired to write guidelines. At the day-to-day policy level nothing happens and the traditional regulatory practices prevail. If this continues, economics turns out to be an ideology covering inefficient policy with the folklore of the market. Unfortunately, the long-term results will be equally damaging for the economy and the environment. Nevertheless, work such as has been achieved by the Intergovernmental Panel on Climate Change (cf. DOKKEN ET AL., 1999) makes me optimistic that future policy will falsify my fears.

References:

Armstrong, H. and Taylor (1993), J., Regional Economics and Policy, Second edition, New York, London Harvester Wheatsheaf

Aring, J., Holst, M., Altena, O., Schnur, O. (1995), Die Bedeutung des Flughafens Hamburg für die Metropolregion Hamburg, Untersuchung im Auftrag der Wirtschaftsbehörde der Freien und Hansestadt Hamburg, Hamburg

Baumol, W.J. and Oates, W.E. (1988), The Theory of Environmental Policy, 2nd edition, Cambridge UP

Beyhoff, S., Ehmer, H., Focke, H. and Warlitzer, V. (1992), Verkehrspolitische Optionen zur Lärmreduktion an Flughäfen dargestellt am Beispiel des Flughafens Hamburg, Deutsche Forschungsanstalt für Luft- und Raumfahrt Forschungsbericht (DLR), Köln

Bonus, H. (1984), Marktwirtschaftliche Instrumente im Umweltschutz, Stuttgart

Bonus, H. (1992), Umweltpolitik in der sozialen Marktwirtschaft, (Manuskript) Fachtagung Umweltökonomie, Mannheim

Button, K.J. and Reynolds-Feighan, A. (1999), An Assessment of the Capacity and Congestion Levels at European Airports, in: Journal of Air Transport Management: Vol. 5: 113-134

European Commission (1998), Consultation Paper on Air Transport and the Environment. European Commission, Directorate General for Transport, Brussels, Belgium July 1998

Flughafen Hamburg GmbH (1996), Antrag der Flughafen Hamburg GmbH auf Planfeststellungsbeschluss gemäß _ 8 Abs. 1 LuftVG für die Änderung des Flughafens Hamburg durch Errichtung und Betrieb eines Vorfeldes 2 nebst damit im Zusammenhang stehenden und im Antrag konkretisierten Maßnahmen und durch eine neue Lärmschutzhalle vom 23. Dezember 1996.

Dokken, D., Griggs, D., Mc Farland, M., Lister, D., Penner, J. (1999), Aviation and the Global Atmosphere, A Special Report of the Intergovernmental

Panel on Climate Change Working Group I and III, Cambridge University Press

Isermann, U., Koppe, E., Mueller, E.-A., Meyer, T.J., (1992) Lärmphysikalisches Gutachten für den Flughafen Hamburg, Untersuchung im Auftrag der Umweltbehörde der Freien und Hansestadt Hamburg, Hamburg

Isermann, U., Piehler, C. Urbatzka, E. (1997) Gutachten zur langfristigen Luftverkehrs- und Fluglärmprognose für den Flughafen Hamburg, DLR Hauptabteilung Verkehrsforschung, Köln Institut für Strömungsmechanik, Göttingen, Untersuchung im Auftrag der Wirtschaftsbehörde der Freien und Hansestadt Hamburg, Hamburg

Kunz, M. (1999), Airport Regulation: The Policy Framework, in Pfaehler, W., Niemeier, H.-M., Mayer, O.G., Airports and Air Traffic: Regulation, Privatisation and Competition, Economic Forum of the Free and Hanseatic City of Hamburg, Frankfurt M., New York, Peter Lang Verlag: 11-55

Leitlinien für eine norddeutsche Luftverkehrspolitik (1995), Beschluss der Konferenz der Wirtschafts- und Verkehrsminister/-senatoren der Norddeutschen Küstenländer vom 12.6.1995, Wirtschaftsbehörde der Freien und Hansestadt Hamburg, Hamburg

Mandel, B. (1997), Luftverkehrsprognose für den Flughafen Hamburg, Untersuchung von MKmetric im Auftrag der Wirtschaftsbehörde der Freien und Hansestadt Hamburg, Hamburg

Mandel, B (1999), The Interdependence of Airport Choice and Air Travel Demand, in Gaudry, M., Mayes, R., (editor), Taking Stock of Air Liberalization, Dordrecht, Kluwer academic Publishers: 189-222

Mishan, E.J. (1984), Cost-Benefit Analysis, 3rd edition, Allen & Unwin London

Mueller, J. (1998), Methoden der Bewertung des wirtschaftlichen Einflusses der Flughäfen, (Manuskript), Berlin

Niemeier, H.-M. (1998), Praktische und theoretische Probleme einer Wirtschaftspolitk für den Flughafen Hamburg, in Priddat, B. P. und Vilks,

A., Wirtschaftswissenschaft und Wirtschaftswirklichkeit, Marburg: 181-206

Pfähler, W. and Gabriel, C.M. (1999), Die regionalwirtschaftliche Bedeutung des DESY in Hamburg. Input-Output-Analyse der Nachfrageeffekte, Hamburg

Rothengatter, W. (1994), Do Benefits Compensate for External Costs, Transport Research-A, Vol. 28: 321-328

Starkie, D and Thompson, D. (1985), Privatising London's Airports, Institute for Fiscal Studies, Report 16, London

Tse, K. and Tretheway, M.W. (1995), A Study of Economic Multipliers and their Applications to the Economic Impact of the Vancouver International Airport, Vancouver, Vancouver International Airport Authority

Wirtschaftsbehörde der Freien und Hansestadt Hamburg (1998), Planfeststellungsbeschluss der Wirtschaftsbehörde der Freien und Hansestadt Hamburg gem. _ 10 Abs. 1 LuftVG in Verbindung mit _ 1 der Zuständigkeitsanordnung des Senates über Zuständigkeiten auf dem Gebiet des Luftverkehrs für die Änderung des Flughafens Hamburg einschließlich der Folgemaßnahmen gemäß den Planunterlagen sowie nach Maßgabe der ergänzenden Bestimmungen nach __ 8-10 LuftVG und _ 72ff HmbVwVfG vom 26.05.1998 (Planfeststellungsbeschluss Flughafen Hamburg für die Erweiterung des Vorfeldes 2 auf 23 Abfertigungspositionen sowie für weitere Infrastrukturmaßnahmen und für den Bau eines Vorfeldes 3 für die allgemeine Luftfahrt).

PART III

ENVIRONMENTAL POLICY OF AIRPORTS IN THE AGE OF PRIVITAZATION

Measuring the Benefits and Costs of Alternative Noise Management Strategies

David Gillen, Professor of Economics and Transportation
Institute for Transportation Studies
(University of California, Berkeley)

Introduction

People have relied on governments to provide social regulation aimed at controlling externalities – spillovers from everyday business activities- that compromise or endanger human health and safety. Typical targets of social regulation include product safety, workplace safety, pharmaceuticals and discharges of air and water pollution. Of course the role of government is appropriate because the externalities reflect failures in markets to correctly (efficiently) price these activities or their outputs and government acts as an efficient arbiter. Recently, however, social regulation is coming under increasing attack as being excessively costly. In a recent paper, Hahn and Litan (1997) estimate that social regulation in the United States costs firms and consumers roughly $200 Billion while producing benefits that fall significantly short of this amount.

A number of politicians and policy analysts have called for regulations to be put in place that would require agencies to conduct a benefit-cost analysis of any regulation that will have an economic impact greater than some fixed amount - $100 million. This appears to be a step in the right direction since the optimal amount of anything, including pollution or safety, is only that which is worth what it costs. Therefore the fundamental economic criterion for evaluating any public policy is whether is *maximizes* net social benefits. Granted that social regulation is designed to reduce the welfare losses caused by externalities or threats to human health and safety, a first step in this process is for regulators to know whether the alleged market failure is sufficiently costly that it warrants government intervention and whether the proposed remedy chosen is the most efficient way to correct the market failure.

The regulation of airport noise provides a good illustration of how addressing these questions can lead to an improvement in regulatory policy by targeting government intervention where it is needed and in selecting the right mechanism when it is. In the battle between airports and their environs, in all parts of the world, the solutions generally selected are to improve technology or to limit exposure through land use planning. For example, the Airport Noise and Capacity Act passed in 1990 in the U.S. has mandated the elimination of Stage

II aircraft at all U.S. airports by the end of 1999. This legislation forces the retirement of approximately 44 percent of airline's Stage II aircraft that would have remained in service past 1999. Even though many of these aircraft can be sold into markets outside of the United States, it still imposes a sizable capital loss on the airline and aircraft leasing industry. At the same time it provides a significant wealth transfer to current homeowners who are in the effected region of airports. One has to ask is the question wrong or the solution wrong? That is, should the government be intervening at all or should it intervene but use a different means of achieving the 'right' amount of noise. In a recent paper, Morrison et al. (1998) argue that the use of an optimal noise tax generates net benefits of only $0.16 Billion (present value). Because the net benefits from efficiently charging for aircraft noise is so small, the main effect of the noise policy is to transfer wealth from the airline industry to homeowners rather than to improve social efficiency. The FAA noise regulations, they argue, have generated substantial costs when there is little justification on efficiency grounds for intervention in the first place.[1]

This paper examines the benefits and costs of alternative noise management strategies. It begins by asking the question should government be intervening at all in this issue and if so how. In section 2, the question of the need for government intervention is addressed. Section 3 contains a decision analysis framework for selecting among the different strategies available while section 4 develops a comprehensive set of noise management strategies. Section 5 provides an outline of the benefits and costs of each noise management strategy and section 6 contains the quantitative measurement of three different impacts associated with noise management. Section 7 contains a summary and some conclusions.

2.0 Is Government Intervention Warranted

One could take the narrow position that economic efficiency is the sole criterion for judging whether and how governments should intervene in markets. Or one can take a broader managerial perspective that recognizes that airport externalities need to be managed and failure to do can result in significant economic losses. This view recognizes that people affected by externalities do not recognize 'economic efficiency' as the sole criterion for judging whether an action is good or bad. Their perspective is a private one and their actions can impact airport operations, efficiency and productivity.

Environmental spillovers from airport operations challenge airport managers. This is especially the case for airport noise. Airport neighbors increasingly

[1] They conclude that if social regulation rests on solid analytical basis, questions of cost-efficiency would not arise.

oppose management's capacity expansion plans in their efforts to contain noise exposure. Their access to formal environmental hearings forces airport management to deal with the conflicting demands of airlines, their customers and the airport's neighbors. Airport managers find themselves "caught in the middle" of those who demand solutions to airport congestion and those who demand solutions to noise.
Effective noise management requires strategies that balance the needs of airport owners, airlines and their customers, and the airport's neighbors.

Externalities create a strategic problem for an airport's manager because the recipients can retaliate. For example, the decision to reschedule movements to one runway while another undergoes repair typically produces increased public complaints about noise. The strategic significance of the externality problem depends directly on the seriousness of the retaliation that might arise. An assessment of the needs and potential for retaliation of the "players" in an airport's environment represents a crucial step in the development of a noise strategy.

At issue in the dispute between an airport's neighbors and its customers is the level of use of a common property resource -the surrounding physical environment. Demands of landowners who face expropriation, complaints about noise, and political opposition to expansion plans influence the feasibility of increasing airport services. Expropriation costs reduce the profitability of expansion investment. Political forces can reduce the probability that regulators allow expansion at all. Both demands, then, are expressions of a " supply price" for a resource that the airport needs.
An airport's customers similarly value the use of the surrounding environment but for different reasons. Their "demand price" is the maximum that they would be willing to pay to use the resources of the airport as a transportation facility. In the absence of a market in which both sides can agree about how noisy airport activity will be, the airport's owners and managers must provide strategies that coordinate the demands of its neighbors and its customers. Otherwise they will face the obstacles and constraints that arise when one side acts on its dissatisfaction.

Airports in North America, Europe and Asia are experiencing extraordinary pressures to accommodate increased traffic by adding more slots and runways. Simultaneously, a general demand for more environmental responsibility is forcing airports to listen to and deal with groups affected adversely by airport externalities. The urgency of the congestion problem coupled with this environmental awareness is increasing the public's strategic leverage in approval processes. This leverage could have a significant negative impact on the rationalization of the worldwide air transportation system. Airport managers must respond by adopting comprehensive noise strategies that balance the demands of their customers and their neighbors.[2]

A common response to noise complaints in the past denied that noise generates costs, and focussed instead on progress and economic benefits derived from more intensive use of airport resources. Nevertheless, noise is costly, and not just as it negatively affects an airport's neighbors. To the extent that these neighbors have any strategic leverage, they can delay or block indefinitely socially efficient increases in airport capacity.

Growing public awareness and acknowledgement of negative externalities like air and water pollution have moved the resolution of externality problems to the front of the political agenda. It is easier today for opponents of airport expansion to use the political system by participating in environmental impact hearings and electoral politics. Even if an airport's managers are not under pressure to expand, they face a greater strategic challenge based on their environmental roles.

Management's "middle ground" role depends on its responsibility to its owners to run their airport as efficiently as possible. Management must develop strategies to deal with spillover problems if they represent potential impediments to the attainment of this objective. An airport's particular circumstance will determine the threat and the appropriate response. Whatever the response, it is best determined by a rational assessment of the problem, including a benefit/cost analysis of alternative strategies.

3.0 A Framework for Decision Analysis

A number of strategies have been developed to reduce noise. Each of these strategies has associated costs and benefits that are borne by different groups. Costs have not been seriously considered nor systematically assessed when choosing from among the various noise strategies available. These costs can be significant and we suspect that the magnitude and incidence varies across the alternative strategies. Among the costs are land use controls which may limit the

---

[2] An airport's neighbors have gained increasing strategic power for several reasons. Many airports' interest in capacity expansion has never been higher. Congestion and unabated growth in demand for air transportation have increased the potential costs of failing to implement successful capacity expansion plans. This makes the threat posed by dissatisfied airport neighbors considerably more important than it would be if the industry were experiencing no growth. Airport managers are responsible for operating an airport in the interests of its owners. Rational management practice requires them to make decisions that can be justified on a benefit/cost basis. Failure to anticipate retaliation and manage accordingly means that the full costs of a decision are unaccounted for, because retaliation can impair management's ability to optimize the use of the airport.

productive use of land as well as tax revenues to municipalities, and operating procedures which may have an impact upon operating and scheduling costs or result in flight delays (due to special routing or corridors). The move to Stage 3 aircraft will impose significant capital costs upon the carriers. Airport closures, night curfews etc. impose costs on the airport in terms of lost revenue and reduce available capacity at a time of already significant congestion. The procedures can also result in under-utilization of aircraft particularly in the case of airfreight companies. In some cases noise fees are established with no real economic basis.

On the other side the noise management strategies will have some impact on the magnitude of the noise externality which airport neighbors are faced with. But what management strategy will have the most effect in reducing the 'noise problem'? This will depend on a number of factors two of which are; which features or characteristics of the noise problem does the management strategy affect and, second how important is each feature to the noise recipients?

In order to *develop an efficient and effective noise management program the magnitude and distribution of the benefits and costs* of the strategies must be considered. Which strategies are more efficient? To what extent should a particular strategy be implemented? Where do diminishing returns set in and where do marginal costs and benefits equalize? Without such estimates not only will airport and air carrier resources be wasted but more harm than good may result from well-intentioned but ill-based strategies aimed at mitigating the noise problem.

3.1 Externalities and Social Cost

Making investment decisions or deciding which management strategies to pursue is, in principle, a straightforward process. The gains from the investment or strategy are compared with the costs. Investments and strategies are ranked according to their return and with the objective of maximizing the rate of return, the decision is self-evident. A similar picture could be developed with respect to a consumer making decisions with regard to purchases; the consumer would rank choices on the basis of costs and gains and select the bundle of goods which maximizes the net gains.

These seemingly simple propositions of conventional economics are based upon the notion that prices reflect consumer desires and in the long run will direct resources where they are most productive. The examples above also mask simplifying characteristics: there is a common goal, profit or utility; costs and benefits accrue to the same economic agent or the weights to be attached to different groups who may be the recipients of benefits or costs can be identified; prices reflect all the costs and gains; and markets are well defined.

Noise is an excellent example of an economic good which is "bad" or undesired. It is a product for which there is no well-defined market and hence no well-defined price. The producers of noise are generally not the consumers of noise; thus, the price of goods for which noise is an undesirable by-product do not reflect the full "social costs" of their production.[3]

How should decisions with respect to noise reduction strategies or investments, such as engine redesign be made in order to mitigate noise?[2] The economics of such a decision demands that first, measures of the gains and losses are made and then an objective standard formulated, then, a decision as to which strategy or set of strategies to use can be made. In order to accomplish this, both costs and benefits must be measured in some commensurate units. This requires "monetizing" noise or valuing noise in terms of a dollar metric or at minimum being able to rate the relative values of different strategies. Costs of particular strategies or investments are *relatively* straightforward and can be appraised. However, the benefits of noise reduction or the valuation of quiet are not easily evaluated.

3.2 Social Costs and Strategy Choice

One way of viewing the effects of various noise management strategies is to consider how they affect costs and benefits. Noise from an aircraft operation is a by-product, or externality, of the output of the aviation industry. Figure 1 illustrates the "optimal" amount of noise. The notion of an optimal amount of any good is that amount which is worth what it costs and in most all cases the optimal amount is not zero.[4] In effect there has to be a balancing of benefits and costs.

In Figure 1, q* is the optimal level of noise (or quiet), and it illustrates that the optimal level occurs where the incremental benefits (MB) of additional noise reduction just equals the incremental costs (MC) of reducing noise. The total benefits of a level of quiet, q*, are measured by the area under the MB function between 0 and q*. The total costs of reducing noise to the level indicated by q* is given by the area under the MC function between 0 and q*.

The efficient level of quiet, q*, is achievable in a number of different ways. Furthermore, q* can be valued in different ways as well because it has multiple dimensions. By this we mean that the noise externality may be described by the cumulative amount of noise, the average amount of noise, the maximum amount of noise at a given moment in time, the frequency of exposure to noise.

---

[3] Airfares do not in most cases contain a price for the use of quiet.

[4] To eliminate or have zero noise from aircraft would require shutting down an airfield. This could be very costly in terms of employment, generated economic activity and development and loss of investment in capital and land.

Figure 1

The Optimal Amount of Noise

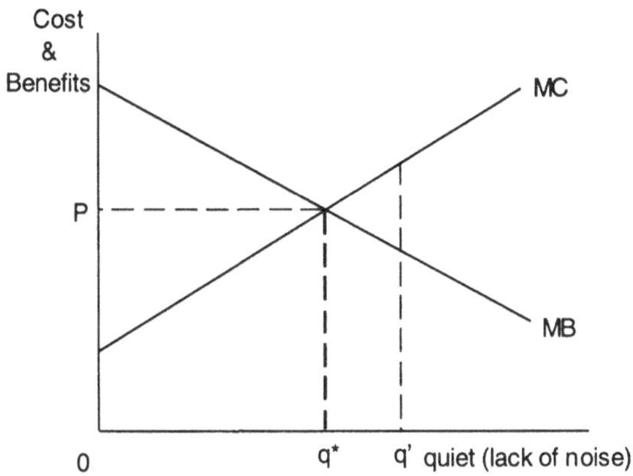

The benefit function is a simple linear combination of these characteristics. Each may have a different value. It is even possible that a particular characteristic has a very low or zero value. Two points are worth noting. First, if the valuation of any one characteristic increases the MB function rises. Thus if airport neighbors become more noise sensitive, MB shifts upward. Second, if a noise strategy is put in place that affects a noise characteristic, which has little or no value, MC will most likely exceed MB. Such a situation is illustrated in Figure 1 at q'. Therefore, Figure 1 can be thought to apply to any one-noise strategy as well as to the aggregate noise problem. Each noise strategy has costs and benefits. To select the 'best' one requires that for each strategy all costs and benefits be measured and compared.

Once benefits and costs of the alternative strategies have been determined the relative value of a strategy can be assessed. To illustrate how even a coarse classification of cost and benefit impacts can assist in sorting noise management strategies, suppose that each noise management strategy is ranked in terms of noise reduction versus noise redistribution with reduction being preferred to redistribution. A strategy which reduced noise could be rated as 'High Benefit' and redistribution would be 'Low Benefit'. Recent research by Gillen & Levesque (1993) provides information by which strategies that reduce noise can be ranked. Specifically, noise reduction from a decrease in noise energy from a given number of aircraft (more Stage III, for example) is valued higher than a noise reduction resulting from a decrease in flight frequency.

From other analysis, the cost of each strategy might be classified as 'low' or 'high.' Figure 2 shows three strategies that have been classified this way. This case illustrates a ranking based on two value categories, 'high' costs and 'low' costs, and high' benefits and 'low' benefits. While the information on costs and benefits is minimal, it is still possible to conclude that the strategy classified as 'high' cost and 'high' benefit improves on the strategy that is 'high' cost and 'low' benefit. Therefore, from even a relatively gross measure of value it may be possible to establish a partial ranking of the strategies. As more precision is added to the values of the costs and benefits, the alternative strategies can be ranked more thoroughly. In effect, with finer valuations, the analysis moves from something like Figure 2 to Figure 3 which is based on a full information cost and benefit analysis. Figure 3 illustrates the application of this framework to the assessment of four alternative noise management strategies. Any strategy which is on or above the 45° line is one which yields a positive return; the dollar value of benefits per dollar of costs is > 1. In the illustration, strategy 4 is one, which should be rejected since it costs more than the benefits received. The economically efficient strategy is strategy 1 since it yields the greatest benefits per dollar of costs. However, all of strategies 1 through 3 yield positive net benefits.

Figure 2

Ranking of Noise Management Strategies

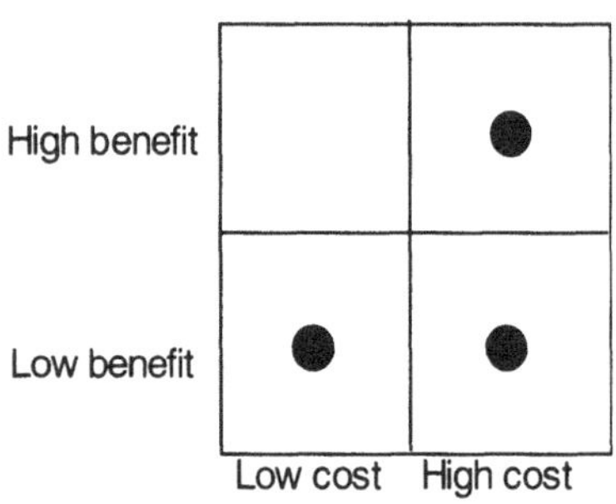

Figure 3

Benefits and Costs of Noise Management Strategies

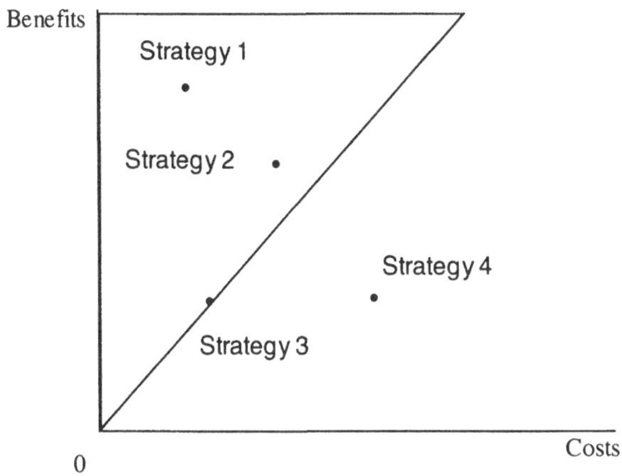

In order to select the most efficient and effective strategy it is necessary to:

1. identify the costs of each strategy
2. assess whether a strategy alters the characteristics of noise which are annoying
3. determine how a reduction in the noise externality is valued
4. determine the population impact of a strategy.

4.0 Noise Management: A Comprehensive View

Noise management affects different groups in different ways and can be more or less effective under different circumstances. The strategies available to airport managers vary in terms of their impacts on the properties of noise. Some affect the loudness of events primarily, some affect the frequency of events, and others affect both. It is important in assessing noise management strategies to decide what the 'noise problem' is, whether noise annoyance has to do mostly with the loudness of events or the number of events. Poorly conceived strategies that do not address the needs of airport neighbors or that assume a nonexistent noise problem waste resources.

There are a number of persons, groups, firms or agencies that have a stake in the airport and in the choice of which noise strategies are selected. The members of each of these interest groups have objectives and values, which may be coincident or conflict with those of the airport. If airport activity affects them negatively it is they who can through the regulatory, policy and political process affect the business of the airport. An airport manager, in evaluating costs and benefits must consider the relative importance of these stakeholders. The gain or loss to each stakeholder from a noise management strategy should be weighted by the relative importance of that stakeholder.

Table 1

Stakeholders in the Strategy Selection

Stakeholder	Preferences, Values, Expectations	Ways of Affecting Airport Objectives
Airlines		
Passengers		
Cargo		
Community		
Airport Neighbors		
Airport Services		
Airport Employees		
Pilots		
Airport Owners		
Policy Makers		

Table 1 provides a suggested list of the interest groups associated with an airport. As illustrated each group have their objectives, expectations and potential impact on airport operations. The purpose of examining each, as we have suggested, is to weight their relative importance in the selection and implementation of the noise strategy. How these groups are viewed will depend very much on the organizational structure and institutional makeup of the airport. Table 1 is a 'worksheet' for a stakeholder analysis. There are numerous measures of noise exposure. Some measure the amount of noise energy, others measure the maximum exposure of noise energy for a particular event or occurrence while still other cumulative noise measures are a composite of noise energy, event frequency and weighting for the timing of the events. These measures are defined in terms of the supply of noise rather than in terms of the recipients' response to it. There is a literature, however, which provides evidence that noise annoyance is highly correlated with these supply measures. The most commonly used measures are the cumulative type, NEF, Ldn, NNI, for example, but these mask the problem because they do not allow identification of which noise characteristic is the essential problem at a particular airport. The noise problem has potentially four dimensions; the number of flights, the location of the flights, the timing of flights, and the sound levels emitted from those flights. Each noise strategy affects one or more of these characteristics. *Task one* of the airport manager is to identify which features of noise characterizes the noise externality and are relatively more important in their case. It is this, which provides the basis of the benefits for a reduction in the externality.

Table 2

Noise Management Strategies

Management Strategy	Noise Characteristics Affected	Benefits (to Whom)	Costs (to Whom)
Airport Planning – Prevention			
Runway Location			
Displaced Thresholds			
Isolate Run-up and Maintenance Noise			
Noise Berms			
Noise Easements			
Avigational Easement			
Taxi way and Runway Extensions			
Airport & Airspace Use			
Preferential Runway System			
Arrival or Departure over Open Space			
Engine Ground Run-up			
Rotational Runway System			
Nav Aid Assisted Departures or Arrivals			
Aircraft Operations			
Maximum Safe Climb on takeoff			
Takeoff Thrust Reduction			
Reverse Thrust Limits after Landing			
Informal Flight Operations Restrictions			
Aircraft Restrictions			
Runway Restriction			
Limit on Operations by time Period			
Limit on Noise Capacity			
Limit on Basis of Noise Levels			

Other			
Purchase Assurance			
Sound Proofing			
Noise User Fees			
Noise Budgets			
Noise Monitoring			
Flight Training Restrictions			
Weight & Thrust Limit			
Restrict by Noise Characteristics			
Airport Curfew			

Table 2 contains a comprehensive list of noise strategies available to the airport manager. They are identified by the noise characteristics they affect, who benefits and who pays the cost. Table 2 is a 'worksheet' for assembling pertinent technical information about noise management strategies. What properties of noise do each affect? Who benefits from the strategies (how many people, where are they located)? Who bears the costs of the strategies?

From the larger list of noise strategies available a subset was investigated and these can in turn be categorized into three generic categories; **airport and airspace use**, **aircraft operations** and **aircraft restrictions**. This categorization represent strategies which are short or intermediate term in terms of implementation and because they can be directly affected by the airport manager.

5.0 Noise Management: A View of the Benefits and Costs

In our analysis of the noise management problem we group noise management strategies into five broad categories and these are illustrated in Table 2. The first, Airport Planning is a longer term set of strategies which the airport can undertake alone and in conjunction with planning authorities in the surrounding community. Most of the strategies in this category, however, require some investment and therefore a longer planning horizon. The next three groups, Airport and Airspace Use, Aircraft Operations and Aircraft Restrictions are shorter term strategies which lie, for the most part, within the purview of the airport manager. In some cases the Federal government must approve strategies if they conflict with policy or agreements within the aviation sector.

Each noise strategy carries with it costs and benefits for one or more of the stakeholders. What has been missing is the identification *and* valuation of these costs and benefits. Previous work has either ignored many categories or has operated at a high level of aggregation. For example, the hedonic literature on housing values and noise impacts provides a gross dollar measure of the impact of changes in some cumulative noise measure. What it has not done is to identify the costs to other parties with the introduction of a noise strategy or the benefits to those same parties if there is an increase in airport activity (or equivalently an increase in the noise externality). Furthermore, no attempt has been made to decide whether one strategy is more economically efficient than another. This is attributable to the lack of detail in the analysis.

One way of addressing this problem is to estimate or calculate a value for each of the various cost and benefit impacts that are associated with a strategy. There entails answering two questions. What are the probable impacts or effects of a noise strategy? Table 3 shows possible impacts associated with three classes of noise strategies. It is speculative because at this point since there is no documented and systematic evidence concerning these cost and benefit impacts. As an initial step toward providing this information, this study examines 'expert' opinion about the effects of various noise strategies. This is a starting point for the more comprehensive application of engineering cost analysis and externality valuation methods.

The second question and step toward a complete NMSA concerns attaching dollar values to the benefit and cost impacts. Having identified the categories of costs and benefits arising from a particular strategy, managers at an airport can study the magnitudes of these impacts for their situation. The categories direct them to the questions and calculations that they must make to assess a strategy.

Table 3

Airport Noise Strategy Impacts

1. Airport and Airspace Use

Costs

- airline operating costs, fuel crew time, flight time, capital depreciation
- reduced airport capacity
- higher ATC costs and lower productivity
- pilot load
- airline scheduling and aircraft positioning
- passenger time: timing of flights and frequency of flights affects desired vs. actual travel time

Benefits

- noise concentrated in corridor - more noise on fewer people
- reduced noise concentration, broadly shared over community
- reduced noise for SENEL and flight frequency
- noise reduction at sensitive times

2. Aircraft Operations

Costs

- increased pilot workload
- reduced airport productivity
- reduced airport capacity
- airline less flexible in use of fleet
- reduce productivity of airline fleet

Benefits

- small reduction in noise
- noise reduction to small portion of community
- small noise reduction over broad community

3. Aircraft Restrictions

Costs

- airline costs of reduced capital use and scheduling
- system efficiency reduced
- ATC productivity reduced to position aircraft
- cargo/courier costs rise with increase in rate of depreciation of Stage II aircraft
- affect competitive advantage between different carriers

Benefits

- significant reduction in noise everywhere (even where it has a low marginal benefit)

As an interim step, some measure of the relative size of the different cost and benefit impacts can provide further direction about the categories that ought to be studied particularly carefully. As an example, managers' should concentrate their measurement efforts on airlines costs if changes in airport and airspace use, such as 'preferential runways', if they understand this to be the largest cost effect. This weighting of the cost impact also provides management with an idea about which stakeholders are most likely to resist a particular strategy.

6.0 Assessing the Costs of Noise Management Strategies

Eventually, a complete NMSA involves estimates of the size of the cost impact associated with a noise management strategy. If aircraft restrictions increase airline-operating costs, engineering cost analysis of the magnitude of these effects is required. The preparation for this level of detail requires some understanding of where to look for cost impacts. This knowledge of the 'technology' of noise management strategy provides a framework for the manager trying to understand the relative merits of different approaches. At the same time it is possible to look to secondary markets to assess the cost (or benefit) impacts of different noise management strategies. Generally the cost or benefit categories will include airline costs, airport costs and community costs.

In section 6.1 a framework is presented that provides a mechanism for ranking the relative effectiveness and efficiency of the different noise management strategies. Sections 6.2, 6.3 and 6.4 provide more detail measurements of particular costs and benefits associated with noise.

6.1 Cost-Benefit Categorization: First Steps

Gillen and Levesque (1990) provide a partial framework for categorizing the costs and benefits of various noise management strategies. Currently, however, there is no documented and comprehensive description of the potential cost impacts of noise management strategies. As well, while the cost estimates for particular strategies will depend on the airport, information concerning the relative size of the different cost categories can be useful in allocating management attention to the effects and in formulating partial assessments of the merits of different strategies.

While there is not a systematically documented understanding of the cost impacts of noise strategies, that knowledge nevertheless exists informally in the community of airport managers who deal with noise. Further, this knowledge probably includes estimates of the relative magnitudes of the cost effects from noise management strategies focusing on: airport and airspace use strategies, aircraft operations strategies, and aircraft restriction strategies. One might assume that these groups of strategies share common cost and benefit impacts (although the magnitudes may differ depending on the specific strategy).

A first attempt was undertaken by Gillen and Levesque (1994) to compile an assessment of the relative values of costs and benefits associated with these listed strategies. They employed an 'opinion thermometer,' a form of graphical scale that can be used to provide estimates of relative magnitudes. Figure 4 illustrates this scale. Table 5 shows the results of one of the tests. A sample of airport managers responded to the survey instrument. They were asked to simply check whether each cost impact listed is associated with the effects of the relevant strategy type.

Figure 4

The Opinion Thermometer

RATING

- fuel costs increase ______
- airport capacity declines ______
- crew costs increase ______
- ATC workload increases ______
- flight times increase ______
- pilot workload increases ______
- maintenance costs increase ______
- scheduling is harder ______
- airlines' capital costs rise ______
- aircraft use is reduced ______
- monitoring costs rise ______

The table shows the means and standard deviations of the thermometer ratings supplied by the respondents. While the standard deviations suggest that the means are not particularly representative (probably because of the low number of respondents), there is nevertheless something of the same pattern of differences in the noise management technology. Table 5 also shows how weights can be estimated that shows the relative importance of each of the cost impacts.

Table 5

Results of Survey

Cost Impact	Airport and Air Space	Aircraft Operations	Aircraft Restrictions
• fuel costs increase	5.6 (3.3)	5.7 (3.7)	2.4 (3.3)
• airport capacity declines	5.0 (3.4)	3.3 (2.9)	5.6 (3.8)
• crew costs increase	2.3 (2.5)	1.3 (1.3)	3.8 (3.6)
• ATC workload increases	3.3 (2.0)	2.2 (2.6)	1.6 (1.9)
• flight times increase	5.2 (2.6)	3.1 (3.4)	1.6 (2.6)
• pilot workload increases	3.0 (2.1)	4.7 (3.6)	1.6 (1.9)
• maintenance costs increase	2.3 (3.4)	4.8 (3.8)	2.2 (2.6)
• scheduling is harder	1.8 (1.5)	1.8 (2.6)	7.9 (1.4)
• airlines' capital costs increase	2.2 (2.7)	3.9 (3.6)	7.6 (1.9)
• aircraft use is reduced	2.6 (3.3)	2.1 (2.9)	8.3 (1.6)
• monitoring costs rise	1.8 (2.3)	1.0 (2.0)	2.9 (3.6)

6.2 Measuring the Costs to Airlines

Airlines buy, lease and sell aircraft on the basis of a strategy that reflects their network and output expansion in competition with other carriers. They will adopt an optimal replacement cycle of aircraft based on their profit maximizing strategies and because aircraft have a finite economic and physical life. Noise regulations that force carriers to alter the replacement cycle will impose a capital loss on carriers.[5] Because aircraft depreciate with age, use and time their value over time will follow some path. With the announcement of a mandatory phase out at a given point in time will alter this path since the capital will become less valuable.

In order to measure the capital cost to the carriers one can develop a cost model based on capital theory. This model is grounded in the idea that aircraft are durable capital that produces revenue over time. Each period the airline incurs a cost in order to maintain this capital. The impact of a noise regulation such as a Stage II phase out is to make this durable capital less valuable – in the US market it would be obsolete.[6]

If one assumes that airlines replace their initial capital stock, K_0, at some rate δ such that $K_t = K_{t-1} = K_0$. In each time period the airline spends an amount $\delta P_K K_0$ in order to maintain the current capital stock, where P_K is the price per unit of capital. If the discount rate is r, the present value of the lifetime capital cost to the airline will be:

$$\int_0^\infty \delta P_K K_0 e^{-rt} dt = \frac{\delta P_K K_0}{r} \qquad (1)$$

The noise regulation will effect the carriers capital costs in the following way. Consider the noise regulation is announced in period t=0 and that all aircraft must meet the new standard in a given number of years. Suppose β is the portion of the capital stock that does not meet the standard at the time of the announcement, so $\beta P_K K_0$ is the initial value of this capital stock. If one assumes airplane capital depreciates at a geometric rate, it is possible to calculate the geometric rate of depreciation, θ, at which most all of the Stage II fleet's ability

[5] It can also have a significant impact on the level and degree of competition among carriers. This stems from the differences in the age of fleets across airlines. Those with older fleets will suffer a larger loss and may not be able to replace all the aircraft that are forced out of service.

[6] In this example I focus on the phase out strategy in the United States and being considered in other jurisdictions. Other noise strategies such as limiting access to Stage II aircraft at certain points in the day also imposes a cost on the carriers since the capital is less mobile and the carriers have less flexibility in deploying it over their network.

to generate revenue has been lost.[7] The remaining value of the planes will be scrap or the price they can command in markets elsewhere in the world where noise regulations are less stringent.[8] The depreciation rate θ is that associated with the noise regulation while the depreciation rate δ is the one that would occur naturally in the market due to market forces. The difference between these two rates is the cost of the noise regulation. In effect what has happened is the regulation has made the capital obsolete before its time.

The approach a carrier would follow is to covert its' fleet at some rate so at the end of each period a portion of the fleet ρ is still Stage II and $(1-\rho)$ is Stage III. At each point in time a portion, Δ, of the remaining Stage II aircraft are converted. There is a loss in each time period of $\Delta\rho P_K K_0 e^{-\Delta t}$. As carriers replace their Stage II aircraft with Stage III aircraft, the value of the Stage II capital stock grows. The cost to the carrier of maintaining its fleet under the noise regulation will be quite different than if the noise regulation were not in place. The capital cost to the carrier in time t is:

$$\Delta\left(\rho e^{-\Delta t}\right)P_K K_0 + \delta\left(1-\rho e^{-\Delta t}\right)P_K K_0 = \left\{\delta + \rho(\Delta-\delta)e^{-\Delta t}\right\}P_K K_0 \tag{2}$$

The present value of the discounted lifetime capital costs to the carrier under regulation is calculated as

$$\int_0^\infty \left\{\delta + \rho(\Delta-\delta)e^{-rt}\right\}P_K K_0 e^{-rt}\,dt = \left[\frac{\delta}{r} + \frac{\rho(\Delta-\delta)}{r+\Delta}\right]P_K K_0 \tag{3}$$

The increase in the airline's present value of heir capital cost due to noise regulation is the difference between the costs with and without the regulation – the airlines replace their fleet faster than they would have otherwise. The cost is measured as,

$$\left[\frac{\delta}{r} + \frac{\rho(\Delta-\delta)}{r+\Delta}\right]P_K K_0 - \frac{\delta P_K K_0}{r} = \frac{(\Delta-\delta)\rho P_K K_0}{r+\Delta} \tag{4}$$

The cost of the regulation is the difference between the regulated and unregulated depreciation rates, Δ and δ, times the capital stock $\rho P_K K_0$ that is

[7] I use the term 'most all' because with a geometric rate of depreciation the value will never equal zero.

[8] Note the values of these aircraft are much lower as they are not allowed to access the US market.

initially in Stage II aircraft adjusted by the factor (r + Δ) to account for discounting and the declining value of the remaining Stage II aircraft over time.

When this model was applied to the U.S. air fleet Morrison et al. (1998) found that the present discounted cost of the noise regulation dealing with phasing out Stage II aircraft by the end of 1999 was in the range of $7.1 Billion to $13.9 Billion.

6.3 Measuring the Cost of Lost Airport Productivity

Just as certain noise regulation imposes a cost on airlines by accelerating the rate of depreciation of the fleet capital, airports also suffer a cost. When noise regulations are put in place or forced upon airport managers these managers have less flexibility in utilizing their airside and terminal capital. Hence the productivity of this capital is reduced. More resources are used to produce a given output than would have been the case in the absence of the regulations.

If one considers the out put of an airport as number of movements of aircraft, or number of people moved or amount of cargo moved or some composite index of these measures, these are produced using capital, labor, energy, materials and other inputs such a public capital (and the environment). This can be represented by a production function (or its duel cost function),

$$Q_{i,t} = \phi(K, L, E, M, P) \qquad (5)$$

where Q is output, K is private capital, L is labor, E is energy, M is materials and P is public capital. The technology, ϕ, defines the technology which mixes the inputs to produce the outputs. When regulations are put in place that alter this production function relationship it can impose costs in the form of lower productivity. This can be illustrated in Figure 6.

In Figure 6, the airport is producing some output or outputs Q using capital and materials inputs K and M, respectively. One can think of materials input being an aggregation of all other inputs in the production process, labor, energy, materials and resources. In the absence of a noise regulation the airport chooses a technology for producing its output and produces along vector 0E and the total cost of producing this output is 0A. Suppose a noise regulation is put in place that effects the relative cost (price) of capital. The airport is forced to economize on this flight capital and produce using technology represented by the vector 0E" which changes the relative proportions of capital and materials. The cost of producing the output using this inferior technology is 0C or equivalently the area ABB'A' is a measure of the cost of the noise restriction.

Another way of viewing this problem is that the use of the airside capital does not reflect the full social costs. As a result airports select the vector-input combination 0E to produce flights or however one measures output and incurs private cost 0A. However, the real price of capital to include all social costs is represented by the relative factor price line CC' which implies the airport should be using a factor input ratio of 0E'. The difference between the private and social costs is the area ABB'A' or the distance AC.

Figure 6

Measuring the Productivity Effects of Noise Regulations

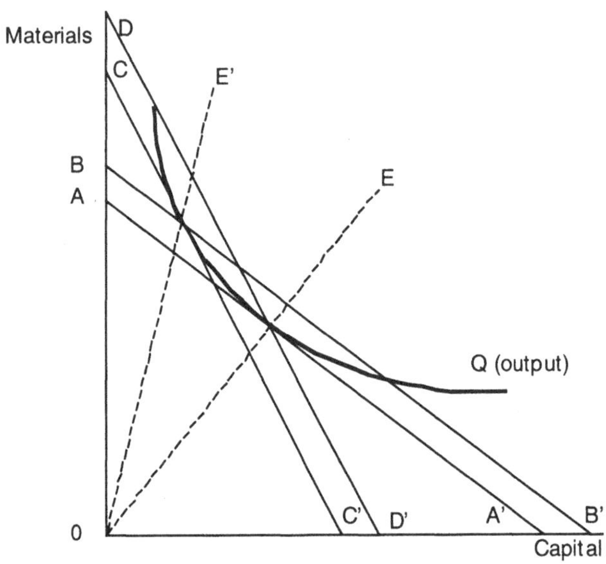

The lost productivity due to the noise regulation or strategy must be measured. To calculate this resource cost it is necessary to generate a measure of airport productivity and then assess the resource cost under a no noise regulation position and those associated with the use of different types of noise management strategies. Gillen and Lall (1997, 1998) have estimated a model measuring airport productivity (distinguishing airside and terminal activities) using Data Envelopment methods and using the calculated productivity measure as a dependent variable in a subsequent regression that assess the productivity impacts of alternative noise management strategies.

In the analysis they develop a database which included the top 30 airports in the United States. The sample uses a subset of 21 of these airports. Information for each airport was assembled for the period 1989-1993 with 114 observations organized in a panel.[9] The first part of the analysis involves deriving a scalar measure of relative efficiency for 114 DMUs.[10] To do this, airports are defined as producing two separate classes of services, these being terminal services and movements. Terminal services are modeled as having two outputs – number of passengers and pounds of cargo and six inputs – number of runways, number of gates, terminal area, number of employees, number of baggage collection belts and number of public parking spots. Movements have two outputs – air carrier movements and commuter movements and four inputs – airport area, number of runways, runway area and number of employees.[11] Movements are assumed to be produced under constant returns to scale (Gillen, 1994) whereas returns to scale are variable in the production of terminal services.

The DEA models used to obtain a scalar measure of relative efficiency are the output-oriented models. The output-oriented model fits in more naturally with the second stage of the analysis, which estimates a Tobit model. Movements are modeled under the assumption of constant returns to scale. The model is as follows for a given DMU:

$$\begin{aligned} \min_{\mu,\nu} g_0 &= \nu' X_0 \\ &s.t. \\ \mu' Y_0 &= 1 \\ -\mu' Y_0 + \nu' X_0 &\geq 0 \\ \mu' &\geq 0 \\ \nu' &\geq 0 \end{aligned} \tag{6}$$

where X_0 is a vector of inputs, Y_0 is a vector of outputs and, μ and ν are weights to be determined by solving the programming problem. Terminal services are modeled using the following variable returns to scale model.

---

[9] Data for Dayton Airport are only available for the period 1989-92.

[10] DMU's are the unit of observation in a DEA method.

[11] Disaggregated data on employees involved in the provision of terminal services and those involved in producing movements were not available. In addition, there were not data on fuel and other inputs such as materials.

$$
\begin{aligned}
\min_{\mu, v, v_0} f_0 &= v' X_0 + v_0 \\
&s.t. \\
\mu' Y_0 &= 1 \\
-\mu' Y_0 + v' X_0 + v_0 \iota &\geq 0 \qquad (1) \\
\mu' &\geq 0 \\
v' &\geq 0 \\
v_0 &\ free
\end{aligned}
$$

where v_0 is the measure of returns to scale or the constant term in the hyperplane and ι is a unit vector. The measure of efficiency reported in Table 6 represents two types of inefficiency, which can be removed by a proportional augmentation of all outputs and residual inefficiency. A value of 1 implies that the DMU is efficient *and efficiency score increases with increase in relative inefficiency.*

An examination of Table 6 reveals the following outcomes for the different airports. In the majority of cases the index is falling implying an increase in efficiency (or reduction in inefficiency). This can reflect in part the emergence from the recession in the late 1980's. However there are some interesting differences. Anchorage for example shows in increase in inefficiency for terminals but a reduction in inefficiency for movements. This reflects the reduction in passengers due to long haul jets being able to bypass this technical stop as well as the growth of hubbing out of the West Coast gateways such as San Francisco and Seattle. Boston's Logan Airport has terminal inefficiency rising and movements inefficiency falling. However, the relative values of efficiency are very different. Terminal efficiency is quite high while airside efficiency was relatively poor but improved markedly over five years.

Table 6

Efficiency Measures from DEA Calculation

Airport	DMU	Terminals	Movements	Airport	DMU	Terminals	Movements
Anchorage	ANC-89	1.0000	2.5554	Memphis	MEM-92	1.0533	1.1728
	ANC-90	1.0947	2.5421		MEM-93	1.0000	1.2105
	ANC-91	1.0333	1.9672	Milwaukee	MKE-89	2.1333	2.3538
	ANC-92	1.1331	1.6731		MKE-90	2.0326	1.9366
	ANC-93	1.1319	1.9505		MKE-91	2.7135	1.8979
Atlanta	ALT-89	1.0658	1.1327		MKE-92	2.5597	1.8847
	ALT-90	1.0000	1.0000		MKE-93	2.5059	1.8513
	ALT-91	1.2093	1.2793	Minneapolis – St. Paul	MSP-89	1.3675	1.8736
	ALT-92	1.1126	1.2635		MSP-90	1.3011	1.7099
	ALT-93	1.0000	1.1856		MSP-91	1.2958	1.8051
Boston	BOS-89	1.1583	1.7050		MSP-92	1.1760	1.6196
	BOS-90	1.1721	1.5258		MSP-93	1.1395	1.3588
	BOS-91	1.2093	1.3533	Ontario	ONT-89	1.0000	2.6907
	BOS-92	1.2014	1.1593		ONT-90	1.0000	2.7001
	BOS-93	1.1311	1.0000		ONT-91	1.0000	2.6846
Baltimore/ Washington	BWI-89	1.6761	1.9109		ONT-92	1.0000	2.7248
	BWI-90	1.6331	1.8677		ONT-93	1.0000	2.3619
	BWI-91	1.6900	1.9422	Phoenix	PHX-89	1.0000	1.0752
	BWI-92	1.7923	2.1235		PHX-90	1.0000	1.0357
	BWI-93	1.9216	1.9613		PHX-91	1.0000	1.0027
Charlotte/ Douglas	CLT-89	1.2254	1.0000		PHX-92	1.0411	1.0000
	CLT-90	1.2980	1.0176		PHX-93	1.0000	1.0274
	CLT-91	1.2657	1.0273	Portland	PDX-89	1.8053	1.2376
	CLT-92	1.1127	1.0402		PDX-90	1.8614	1.1867
	CLT-93	1.1447	1.0000		PDX-91	1.8185	1.1678
Chicago	MDW-89	1.0000	1.0000		PDX-92	1.7312	1.0861
	MDW-90	1.0000	1.0000		PDX-93	1.5044	1.0085
	MDW-91	1.0000	1.1251	St. Louis	STL-89	1.1029	1.8893
	MDW-	1.5640	2.4514		STL-90	1.0860	1.7998

	92						
	MDW-93	1.0000	2.7807		STL-91	1.1371	1.8077
Cincinnati/ Northern Kentucky	CVG-89	1.4838	1.1329		STL-92	1.0572	1.7895
	CVG-90	1.5269	1.0000		STL-93	1.1073	1.8025
	CVG-91	1.4059	1.0000	Salt Lake City	SLC-89	1.4605	2.0072
	CVG-92	1.2417	1.0216		SLC-90	1.3358	2.1787
	CVG-93	1.4439	1.0000		SLC-91	1.3085	2.1127
Cleveland Int'l	CLE-89	1.9690	1.8371		SLC-92	1.2312	2.0320
	CLE-90	1.8451	1.6372		SLC-93	1.0912	2.0392
	CLE-91	1.9278	1.7012	San Diego	SAN-89	1.0000	1.0000
	CLE-92	1.5158	1.4295		SAN-90	1.0000	1.0000
	CLE-93	1.4377	1.6833		SAN-91	1.0111	1.0098
Dayton	DAY-89	1.0000	1.9072		SAN-92	1.0099	1.0000
	DAY-90	1.0000	1.9623		SAN-93	1.0000	1.0000
	DAY-91	1.1825	1.7849	San Francisco	SFO-89	1.0505	2.1147
	DAY-92	1.3434	2.4312		SFO-90	1.0800	2.1741
Fort Lauder-dale	FLL-89	1.7584	1.0000		SFO-91	1.0381	2.1026
	FLL-90	1.6486	1.8308		SFO-92	1.0205	2.2579
	FLL-91	1.7950	1.9414		SFO-93	1.0000	2.4530
	FLL-92	1.7562	2.2228	San Jose	SJC-89	1.0000	2.4739
	FLL-93	1.6305	1.8133		SJC-90	1.0000	1.8111
Kansas City	MCI-89	1.6473	1.5304		SJC-91	1.4144	1.6095
	MCI-90	1.9107	2.1834		SJC-92	1.5807	1.5900
	MCI-91	2.2568	2.6458		SJC-93	1.5957	1.6722
	MCI-92	2.2317	2.3277	Seattle - Tacoma	SEA-89	1.1407	1.0729
	MCI-93	2.8372	2.2539		SEA-90	1.0677	1.0000
Memphis	MEM-89	1.0000	1.5758		SEA-91	1.0000	1.0209
	MEM-90	1.0385	1.7755		SEA-92	1.0000	1.0126
	MEM-91	1.1009	1.6779		SEA-93	1.0000	1.0632

As is true of Salt Lake City but in most of cases the inefficiencies are moving in the same direction. The hub airports exhibit some loose similarities. It appears as hubbing rises the terminal efficiency improves but movements efficiency decreases most likely due to the feed from small commuter aircraft. San Francisco exhibits this result but the level of inefficiency on the movements side is quite startling. San Jose illustrates the impact of American developing their hub there with terminal efficiency rising and movement efficiency gradually falling.

At this point we have addressed only the first part of our questions. We have a measure of relative efficiency and we can rank airports by year and to each other by airside as well as terminal efficiency. This is one of the significant benefits of the DEA measure in that it is useful for both inter-temporal as well as cross-sectional comparisons. However, it is also our purpose to explain the variation in performance measures both over time and across airports. We are interested in two issues. What proportion of the variation in efficiency can be explained by the managerial variables and which noise management strategy variables are most important in affecting this proportion of efficiency?

Table 7 reports the Tobit regression results for movements (airside). The marginal and conditional marginal effects are reported in the last two columns respectively. The two measures are quite similar but the conditional marginal effect is always less. In the movements efficiency regression we are trying to determine which managerial and non-managerial variables have the most impact on the efficiency with which airside can yield aircraft movements. Since the efficiency index has a lower limit of 1, a higher number implies more *inefficiency*. Once you take natural logs the lower limit is zero. So a *positive coefficient implies inefficiency* is going up.

The noise strategy variables are all significant save for limiting operations. Using a preferential flight path or limiting the hours of operation reduces inefficiency while all the remaining noise management strategies tend to increase inefficiency.

Table 7

Tobit Regression on Efficiency related to Movements

Variable	Regression Coefficient	t Statistic	Marginal Effect	Conditional Marginal Effect
Number of Airline Hubs	-0.2725	-4.40	-0.2315	-0.2158
Sq. Ft. of Airside Capacity	0.0001	6.05	0.0001	0.0001
Number of Runways	0.0386	0.99	0.0328	0.0305
Total Number of Gates	-0.0091	-3.03	-0.0077	-0.0072
Proportion of GA Movements	2.8554	8.30	2.4258	2.2609
Year 1989 Dummy	-0.0491	-0.81	-0.0417	-0.0389
Year 1990 Dummy	-0.0431	-0.72	-0.0366	-0.0341
Year 1991 Dummy	-0.0045	-0.08	-0.0038	-0.0035
Year 1992 Dummy	-0.0118	-0.20	-0.0100	-0.0093
Residual Financing	-0.3394	-3.84	-0.2883	-0.2687
Rotational Runways	0.3864	3.33	0.3282	0.3059
Preferential Flight Path	-0.2351	-2.90	-0.1997	-0.1861
Preferential Runway Use	0.2761	2.57	0.2345	0.2186
Limit on Operations	0.0717	0.31	0.0609	0.0568
Limit on Stage II Aircraft	0.6970	3.82	0.5921	0.5519
Limit on Operating Hours	-1.1286	-4.01	-0.9588	-0.8936
Noise Budget	2.0797	5.50	1.7667	1.6467
Atlanta Dummy	1.9590	4.63	1.6643	1.5512
San Francisco Dummy	0.5089	3.11	0.4323	0.4029
Minn. & St. Paul Dummy	2.1189	5.94	1.8000	1.6777
Seattle-Tacoma Dummy	-0.5997	-0.97	-0.5094	-0.4748
Phoenix Dummy	-0.1389	-0.57	-0.1180	-0.1100

Dependent Variable Log of Efficiency Index of Movements
Log of the Likelihood Function 12.390
Estimate of Sigma 0.1833
Asymptotic t-ratio for estimate of sigma 12.986

It appears counterintuitive that limiting the hours an airport is open could reduce inefficiency. However, two explanations are offered. First, if limits on hours and primary hubs are correlated this could explain the result. Airlines will not bring complexes into the airport in the middle of the night to hub since there would be little demand. The second explanation is that given limitations on night use the airport utilizes its' resources very efficiently while those without such constraints do not feel compelled to operate with such efficiency. The preferred noise management strategy is to in fact limit operating hours and to use preferential flight paths. The limitation on operations is clearly counter intuitive except it may be correlated with hub airports. One could also find those airports which have limited operating hours tend to be forced to be more efficient in the time available. It appears the worst noise strategy is a noise budget, limiting Stage II aircraft and using runway rotations.

This method offers a way of including the productivity effects in a benefit-cost calculation for an assessment of noise regulations or strategies. An extension of this work would be to use the efficiency index from the DEA measures or a TFP calculation that would be used in cost function regressions. The inclusion of the efficiency index would tell how much the cost function is shifted due to among other things the particular noise management strategy.

6.4 Measuring the Benefits of Noise Management Strategies

The other half of the NMSA calculation concerns the benefits alleged to flow to airport neighbors from noise management strategies. The survey of airport managers attempts to assess their estimation of these effects. However, analysis of the benefits of noise management strategies properly requires either direct or indirect information from those affected. Hedonic analysis of housing prices is an indirect method that rests on decomposing the price of housing into the quality contribution associated with noise or its components.

The connection between cumulative measures of airport noise and real estate prices has been explored for a number of North American and European airports. Nelson (1980) summarizes the evidence concerning the price effects of airport noise from a variety of studies. He compares their results by denominating the noise effect in a common unit, the Noise Depreciation Index (NDI) (Pearce, 1978). The NDI measures the percentage depreciation for each unit increase in noise exposure. The NDI for 18 cities that Nelson discusses ranges from 0.29 to 1.10, and the average NDI is 0.58. The coefficient of variation is about 0.31, suggesting that the average NDI is quite representative of housing markets in many different cities. Nelson's study and those he cites consider housing markets in the period 1967-80. Pennington, Topham and Ward (1990) and Uyeno, Hamilton and Biggs (1993) examine more recent market data. The former find no statistically significant relationship between housing prices and the Noise and Number Index for Manchester International Airport in the period 1985-86. The latter find that the NDI for Vancouver International Airport is 0.65 for single detached housing in 1987. An unpublished hedonic study by Gillen and Levesque (1990) estimates an NDI of 0.48 for single detached housing neighboring Toronto's Pearson International Airport. In a 1991 study of the housing market around Pearson International Airport Gillen and Levesque (1992) found for single family housing, a one percent increase in the NEF measure reduces housing values *all else held constant* by an average 0.26 percent.

The following table shows the depreciation associated with an increase of 5 NEF for a house of average value for the market ($187,500) and for a variety of initial NEF values. The second column shows what percentage of the initial NEF the increase of 5 represents.

Changes in Property Value with Changes in NEF

NEF	% Change	Lower Bound	Average	Upper Bound
15	33.33%	$14,283	$16,731	$19,202
20	25.00%	$10,712	$12,548	$14,402
25	20.00%	$8,570	$10,038	$11,521
30	16.67%	$7,141	$8,365	$9,601
35	14.29%	$6,121	$7,170	$8,230
40	12.50%	$5,356	$6,274	$7,201

The "Average" column is the result of the calculation $187,500x(0.26788x%Change). The upper and lower bounds represent the 95 percent confidence interval for the estimate. The declining value of quiet as NEF increases is a characteristic of the log-linear functional form.[12]
Calculations of noise effects on residential housing markets exhibit considerable similarity in a variety of cities, at different times. Aviation noise appears to reduce prices of otherwise similar houses between 0.5 and 0.6 percent for each NEF decibel. However, because the NEF measure is cumulative these results do not reveal anything directly about the individual effects of loudness and number of events, information that would be useful in examining the benefits of alternative noise management strategies. A better approach to investigate the loudness and frequency effects involves allowing the two to be estimated independently. The results can then be compared with those based on NEF to determine which is the statistically superior model of noise effects in housing markets.

Gillen and Levesque (1993) undertook to decompose the impact of noise into noise energy and flight frequency. They found that for a "standard" house comprising a 35-year-old three-bedroom bungalow with one bathroom, a fireplace, and garage space for one car. The house is a half-kilometer from the nearest school, and is located on a property of average dimensions. The NDI for average EPNL is about -1.3 over the range 78 to 92; increasing the average EPNL by one decibel reduces predicted house prices by about 1.3 percent.

---

[12] We believe this is a realistic representation of the value of quiet because as long as information flows about the noisiness of neighborhoods are relatively efficient, and real estate markets operate freely, home owners will distribute themselves in relation to noise exposure so that those who are least bothered are closest to the source.

In contrast, the NDI for the number of events is much smaller as would be expected. Adding one more landing or takeoff is less noticeable than raising average loudness. The NDI for the number of events varies from -0.2 to -0.1 as the number of events increases from 80 to 400. The NDI for the variability of the EPNL shows the percentage change in the predicted price arising from a one-decibel *decrease* in the standard deviation. Therefore, in general it appears that relatively noise energy changes are six to thirteen times more important than flight frequency changes. These results carry interesting implications about management of airport noise. For instance, because the number of flights is less important than the loudness and variability of the loudness of single events, the phase in of the quieter Stage III aircraft can be expected to improve overall well being and offset increases in the frequency of events at the same time.

The primary role for the results of the hedonic analysis is in the valuation of benefits. In principle they provide dollar measures of benefits of reducing the exposure of airport neighbors to noise energy or flight frequency. How might these be applied? Any airport manager who is going to use this methodology to develop measures of benefits and costs of alternative noise management strategies must assume that the values calculated from the studies cited are applicable to their airport circumstances or be willing to undertake such an analysis. This would require collecting substantial amounts of housing data and undertaking the requisite statistical analysis. These data are generally obtained from local real estate boards (MLS listings) and such data can be very difficult to secure. A second, and perhaps more important issue to contend with, is the assumption that the values from the housing market analysis are an accurate reflection of what the noise problem is; that is, correctly represent the externality. In a 1982 study of 3580 neighbors of Australian airports Hede and Bullen (1982) find that more than half state that "the steady build-up of all the planes [they] hear in a day" affects their feelings about noise at least as much as the "noisiest plane that [they] hear." Stoddard (1986) estimated a regression model in which the proportion of people 'disturbed' by noise while engaged in activities ranging from sleeping to outdoor relaxation were regressed on several explanatory variables. They included the single-event noise level (SENEL), frequency of day flights, frequency of night flights and income of the individual. The results clearly indicate that (the difference between energy level and ambient noise) in the day/evening noise energy is *three times more important* than frequency. The coefficients can be expressed as the weights discussed above.

Corroborating evidence is found in a recent study from Minneapolis-St. Paul airport.[13] The separate influences of Ldn and the frequency of overflights on noise-related stress were investigated in a regression framework. Four important findings emerged:

- noise energy and flight frequency have independent and statistically different effects on noise-related stress
- the impacts vary by day and season
- the impacts vary by community
- statistically significant differences exist between arrivals and departures

As discussed above hedonic analysis may have limited application in some settings because of an inability to obtain appropriate data to estimate a hedonic price function. This difficulty is partially overcome because the results across settings appear to be fairly portable to others. As discussed above, there is considerable uniformity in the estimates of the price effect of airport noise at different airports and over time. This finding shows that it is probably valid to estimate the 'gross' effects of noise reduction from the results of previous hedonic price studies.
However, it is not possible to reliably decompose the benefits of noise management strategies into those flowing from reduced loudness and those arising from less frequent exposure using most of these studies. Gillen and Levesque (1993) is the only hedonic study that provides any results that might be useful in this connection. Consequently, either more hedonic studies that emphasize noise as a multidimensional phenomenon must be carried out to validate those findings or an alternative and perhaps complementary approach has to be devised to estimate the potential benefits of managing the different aspects of airport noise.

The problem is to design a valid and reliable estimate of the compensation required to offset the effects of different aspects of airport noise. One approach that appears to hold promise involves an application of stated preference modeling. This technique is familiar to transportation economists and engineers who have used it to estimate the demand for alternative travel modes depending on their characteristics, including price, travel time, and convenience.

The literature on noise annoyance documents a variety of activities that neighbors of airports find are negatively affected by airport noise. Bullen and Hede (1982) find that conversation, sleep, television watching, entertaining, and

[13] Airport Noise Analysis, MSP Airport Adequacy Study for the Metropolitan Council of the Twin Cities, James F. Hickling Management Consultants (1989)

outdoor leisure are named by those affected by airport noise as activities that are disturbed. Noise disturbance depends on several factors. Personal characteristics affect an individual's overall reaction to noise; an individual's previous experience with noise, their expectations about noise and their attitudes toward the airport all play a role. These differences mean that the tendency to be disturbed by a given amount of noise will be a distributed variable.

Importantly for the development of a convenient alternative to hedonic price analysis, noise disturbance also depends on the characteristics of noise, how loud it is on average, how frequently it arises, and how variable the events are. These are all properties of noise at a particular location. An individual's choice of housing location may then be seen as depending partially on the noise characteristics at that location and others. These ideas are familiar because they also form the theoretical underpinnings of hedonic price analysis. At this point, however, rather than focus on the market price function, this study turns to the consumer's choice of locations.

The theory of consumer discrete choice shows how to model the probability that a consumer will choose a particular alternative from among a finite set. In particular it models the connection between the probability and the individual's utility for the alternative, which in turn is related to the alternative's characteristics. When this model of choice is combined with an appropriate error theory for the observed choices of consumers, the result is an empirical model that can reveal information about the individual's utility function. The individual's 'value' of noise can be estimated in principle from a sample of observed choices.

To estimate individuals' values of noise, which are expected to differ depending on personal traits, requires several observations. Typically, however, consumers purchase or rent such a small number of housing locations, so infrequently, that there is not sufficient actual data to estimate the desired parameters. An alternative is to have a sample of individuals state what they would do in a variety of hypothetical choice situations that are designed to reveal information about the value of noise. Respondents' information about the alternatives from which they can choose is experimentally manipulated to force them to make tradeoffs between the price of a house and the amount of noise they can expect to experience. In principle, their stated choices in the experiment can reveal information about the respondents' willingness to pay a higher price for a quieter location.

In this approach, because stated choices take the place of actual choices, there is some risk that the results and noise values derived will not be valid because consumers' statements do not match their actual behavior. This risk can be minimized if the experiment is carefully designed to be simple and to realistically draw out a few of the important relationships.

7. Summary

This paper describes methods for framing the assessment of alternative noise management strategies. It is placed in the context of modern airport management and specifically takes the position that noise management is a key component of efficient airport management and noise strategy selection should be based on sound analytical techniques. It outlines a framework that airport managers could use to assess the benefits and costs of a noise management strategy. They could, in addition, use the framework to compare the incremental gains and costs of the different strategies to select from the range of alternatives. Any airport manager should also be interested in the incidence of the costs and the benefits of noise management strategies. Overall, airport management needs to ensure that when airport users bear the costs of the noise strategy that others benefit effectively from the reduction in the noise externality resulting from the implementation of the strategy and by more than enough to offset or equal the costs.

In developing a range of noise management strategies and evaluating them the airport manager needs to:

• identify all cost items that would accrue to an airport operator and residents of communities adjacent to the airport as a result of the imposition of an individual noise restriction;

• identify those cost elements that generically can be attributed to specific (individual) types of noise restrictions;

• develop a methodology including appropriate indicators for use in determining the forecast noise exposure impact reduction associated with a restriction;

• develop a generic cost in terms of housing value change by noise contour number change that results from a restriction;

• develop a methodology and criteria for assessing the impacts of individual noise restrictions.

The work calls for a technical analysis of the cost and benefit characteristics of specific noise management strategies. It also requires an approach for calculating the impacts of a noise management strategy on the pertinent characteristics of noise. These impacts are then to be related to economic values that measure the costs of noise or, the benefits of noise reduction.

The preferred approach is to utilize a benefit cost framework, which identifies for several stakeholders using or affected by the airport, the cost categories for each of several noise management strategies. At the same time one must establish a measure of the gain as reflected in a reduction in the amount of the externality. In all cases one should endeavor to provide a qualitative and quantitative assessment of the magnitude of the cost or benefit and the incidence which the noise strategy has. Using this approach airport managers should be able to evaluate the incremental gains and losses to each stakeholder which results from adjusting an existing noise strategy or with the introduction of a new strategy.
The following steps constitute a formula for an effective and efficient Noise Management Strategy Assessment (NMSA).

• The "noise" problem arises only because someone is bothered by the properties of aircraft operations in their vicinity. It is important to determine the properties of noise that contribute most to the noise problem. These give an initial clue to the likely success of different noise management strategies. For example, the Social Impact Analysis for the Airside Development Project at LBPIA found that 75% of those bothered enough to want to complain were annoyed by flights directly over their homes. In contrast, only eight percent identified the sound of aircraft at the airport as bothersome enough to make them want to complain.

• The range of strategies that managers have available to manage noise is wide. Some simplification and classification of the various strategies is possible by recognizing common technical requirements and common effects on the properties of noise. For example, all strategies that change the geometry of the airport's use result in new flight paths or utilization rates and therefore redistribute the effects of noise. These strategies can be treated as a group in the preparation of an NMSA. It is at this step that the NMSA connects strategies to their impacts on the pertinent properties of noise. Strategies that do not affect the troublesome aspects of aircraft operations can be culled from the list of possibilities at this step; they are likely to be low benefit strategies.

• A strategy identified by its technical requirements and noise impacts, must finally be valued in terms of the benefits that it produces. This entails assigning a money value. An approach familiar to economists involves estimating the contribution of noise measured by the Noise Exposure Forecast to housing price differentials between noisy and quiet neighborhoods. Evidence from a number of studies indicates that representing noise as a cumulative energy measure in house price studies may not be the best way to capture the potential value of noise management strategies effects. This step considers how the information in housing prices might be more effectively used to assess the economic benefits of noise management strategies. Also, because it is not always possible to estimate the money value of noise reduction by this technique, an alternative that elicits equivalent information from surveys is devised.

• The range of stakeholders, those with direct interests in the consequences of noise management, must be developed. Their specific interest in noise management must be documented to get some idea of the cost consequences of strategies. In this step, potential costs are identified for later estimation. The reasoning is of course that in any trade negotiation, and this is trade, distribution matters. It is not enough to simply assess the efficiency consequences.

• The *relative weight* of each cost component identified in the previous step may be more easily calculated than the exact money costs. Initially, before any attempt to estimate engineering costs of noise management strategies, the relative weights assigned by each stakeholder can provide some insight about the overall consequences of a strategy for a stakeholder. For example, the number of stakeholders who identify a cost as very important is a clue to the overall magnitude of the economic equivalent. A survey of stakeholders will provide information about costs from stakeholders' perspective. This is not at odds with conventional economic wisdom. It is simply recognizing there may be more than one price in the market.

• The final step involves amalgamating quantitative and qualitative information about costs and benefits into a characterization of available noise management strategies.

References

Amemiya, T. (1984), "Tobit Models: A Survey," *Journal of Econometrics*, 24, 3-61

Bayless, Mark (1982), "Measuring the Benefits of Air Quality Improvements: A Hedonic Salary Approach", *Journal of Environmental Economics and Management*, 9:81-99.

Bishop, R. and L. Heberlein (1979), "Measuring Values of Extramarket Goods: Are Indirect Measures Biased?", *American Journal of Agricultural Economics*, 61(5): 1183-1205.

Brookshire, David S., Ralph C. d'Ange, William D. Schulze and Mark A.Thayer (1982), "Valuing Public Goods: A Comparison of Survey and Hedonic Approaches", *American Economic Review*, 72:1, 165-177.

Brown, Gardner, Jr. and Robert Mendelsohn (1981), Hedonic Demand Functions with Linear and Non-Linear Budget Constraints, Working Paper, University of Washington.

Bullen, R. B. and A. J. Hede (1982), *Aircraft Noise in Australia: a Survey of Community Reaction*, National Acoustic Laboratories, Commonwealth Department of Health, Report no. 88, Australian Government Publishing Service.

____________________ and E. Kyriacos (1986), "Reaction to Aircraft Noise in Residential Areas around Australian Airports", *Journal of Sound and Vibration*, 108:2, 199-225.

Charnes, Abraham et al., (1994) *Data Envelopment Analysis: Theory, Methodology and Applications*, Kluwer Academic Publishers, Boston.

Cooper, Douglas, and David Gillen (1994) "Measuring Airport Efficiency and Effectiveness in the California Aviation System," Research Report no. UCB-ITS-WP-94, Air Transportation research Center, Institute of Transportation Studies, University of California, Berkeley.

Cummings, R.G., S. Schultze, S. Gerking and D. Brookshire (1982), "Measuring the Elasticity of Substitution of Wages for Municipal Infrastructure: A Comparison of the Survey and Wage Hedonic Approaches", (manuscript, Department of Economics, University of New Mexico, Albuquerque).

Desvousges, W., V. Kerry Smith and McGivney (1983),"A Comparison of Alternative Approaches for Estimating Recreation and Related Benefits of Water Quality Improvements" EPA-230-05-83-001 (March).

Førsund, Finn R., and Erik Hernaes (1994) "A Comparative Analysis of Ferry Transport in Norway," Chapter 15 in Charnes, Abraham et al., eds., *Data Envelopment Analysis: Theory, Methodology and Applications*, Kluwer Academic Publishers, Boston.

Frankel, M. (1988), The Effects of Aircraft Noise and Airport Activity on Residential Property Values: A Survey Study BEBR Faculty Working Paper No. 1450, College of Commerce and Business Administration, University of Illinois, Urbana-Champaign.

Freeman, A. Myrick, III (1978), Hedonic Prices, Property Values and Measuring Environmental Benefits: A Survey of the Issues, Discussion Paper D-37, Resources for the Future, Washington,D.C.

Freeman, A. Myrick, III (1974), "On Estimating Air Pollution Control Benefits from Land Value Studies", *Journal of Environmental Economics and Management*, 1:74-83.

Freeman, A. Myrick, III (1979), *The Benefits of Environmental Improvement: Theory and Practice,* Baltimore:John Hopkins University Press.

Freeman, A. Myrick, III (1981), "On Measuring Public Goods Demand From Market Data", in V. Kerry Smith, ed., *Advances in Applied Microeconomics* Vol. 1, Greenwich, Conn.: JAI Press, pp. 13-29.

Gillen, David and Ashish Lall, "Non-Parametric Measures of Efficiency at US Airport " (paper presented at the ATRG Conference, Dublin, July 1998)

Gillen, David and Ashish Lall, "Developing Measures of Airport Productivity and Performance: An Application of Data Envelopement Analysis ", *Transportation Research E*, (1997)

Gillen, D. and T.J. Levesque, *Noise Management Strategies: A Survey*, Technical Report 89-01 to Transportation Development Centre, Montreal 1989

Gillen, D. and T.J. Levesque, *Alternative Methodologies to Assess and Evaluate the Impact of Aircraft Noise*, Technical Report 89-02 to Transportation Development Centre, Montreal 1989

Gillen, D. and T.J. Levesque, *Assessing the Impact of Aircraft Noise Using the Hedonic Methodology: A Case Study of Toronto International Airport,* Technical Report 89-03 to Transportation Development Centre, Montreal 1989

Gillen, D. and T.J. Levesque, *Using Conjoint Methodology to Evaluate the Impact of Noise and Develop Noise Management Strategies*, Technical Report 89-04 to Transportation Development Centre, Montreal 1989

Gillen, D. and T.J. Levesque, *The Management of Airport Noise*, Final Report to Transportation Development Centre July 1990 229 pages

Gillen, D. T.J. Levesque and T. Smith, *Airport Externalities and Noise Management Strategies in Canada*, Report to Transportation Development Centre, Montreal July 1990

Greene, William H. (1997) *Econometric Analysis*, 3rd Edition, Prentice Hall, New Jersey.

Hahn, Robert and Robert Litan, *Improving Regulatory Accountability*, American Enterprise Institute and Brookings Institution, Washington, D.C. 1997

Hall, Fred, Susan E. Birnie, F. Martin Taylor, and John E. Palmer (1981), "Direct Comparison of Community Response to Road Traffic and to Air Traffic Noise, *Journal of the American Acoustical Society of America*, 70:1690 - 98

Harris, Anthony H. (1981), "Hedonic Technique and Valuation of Environmental Quality", in V. Kerry Smith, ed., *Advances in Applied Microeconomics*, Vol. 1, Greenwich, Conn.: JAI Press pp. 31-49.

Harrison, David Jr. and Daniel L. Rubinfeld (1978), "Hedonic Housing Prices and the Demand for Clean Air", *Journal of Environmental Economics and Management*, 5:81-102.

Harrison, David Jr. and Rubinfeld, Daniel L. (1978), "The Distribution of Benefits from Improvements in Urban Air Quality", *Journal of Environmental Economics and Management*, 5:313-332.

Loehman, E. et al. (1979),"Distributional Analysis of Regional Benefits and Cost of Air Quality Control", *Journal of Environmental Economics and Management*, 6: 222-243.

McMillan, M., B. Reid and David.W. Gillen (1980), "An Extension of the Hedonic Approach for Estimating the Value of Quiet", *Land Economics*, 56, No. 3, 315-328.

Mieszkowski, P. and A. Saper (1978), "An Estimate of the Effects of Airport Noise on Property Values", *Journal of Urban Economics*, 5, 425-440.

Morrison, S., C. Winston and T. Watson, "Fundamental Flaws of Social Regulation: The Case of Airplane Noise" (mimeo, Brookings Institution, 1998)

Nelson, Jon P. (1980), "Airports and Property Values", *Journal of Transport Economics and Policy*,January, 37-52.

Nelson, Jon P. (1981), "Three Mile Island and Residen-tial Property Values: Empirical Analysis andPolicy Implications", *Land Economics*, 57: 364-372.

"Noise Annoyance Costs", James F. Hickling (1990) mimeo

O'Byrne P. et al. (1983), Housing Values, Census Estimates, Disequilibrium and the Cost of Airport Noise: A Case Study of Atlanta, Mimeograph.

Obeng, Kofi, Nasir Assar and Julian Benjamin (1992), "Total Factor Productivity In Transit Systems: 1983- 1988", *Transportation Research A*, Volume 26A, No. 3, pp. 447-455.

Pearce, D.W. (1978), *The Valuation of Social Cost*, G. Allen & Unwin, London.

Pennington, G., N. Topham, and R. Ward (1990): "Aircraft Noise and Residential Property Values Adjacent to Manchester International Airport". *Journal of Transport Economics and Policy,* 24, pp. 49-59.

Phipps, A.G. (1983),"Utility Function Switching During Residential Search", *Geografiska Annaler*, 65B: 1.

Pollakowski, Henry O. (1978), The Choice of Residential Location, Discussion Paper 78-7, Institute for Economic Research, University of Washington.

Ridker, R.G. and Henning, J.A. (1967), "The Determinants of Residential Property Values with Special Reference to Air Pollution", *Review of Economics and Statistics*, 49:246-257.

Rosen, Sherwin (1974), "Hedonic Prices and Implicit Markets: Product Differentiation in Pure Competition", *Journal of Political Economy*, 82:34-55.

Rowe, R. and L.G. Chestnut (1983), "Valuing Environmental Commodities: Revisited", *Land Economics*, 59(40): 404-410.

Sinden, J.A. and A.C. Worrell (1979), *Unpriced Values: Decisions without Market Prices*, John Wiley & Sons, New York.

Small, K. and Rosen H., (1981), "Applied Welfare Economics with Discrete Choice Models", *Econometrica*, Vol. 49, No. 1, pp. 105-130.

Stoddard, Albert T. (1984), "Airport Noise Impact Prediction and Measurement", Ph.D. Dissertation, Cornell University, Ithaca, NY.

Stoddard, Albert T. and Gordon P. Fisher (1984), "A Disaggregate Noise Interference Model for Estimating Airport Noise Impact", *Transportation Research Record* 1033, pp. 47-54.

U.S. Department of Transportation (1985), *Aviation Noise Effects*, Report No. FAA-EE-85-2, Washington, D.C.

Waller, R.A. (1970), "Environmental Quality, Its Measurement and Control", *Regional Studies*, 5, 219-248.

Waller, R.A. (1974),"The Assessment of Environmental Standards" in J. Coppock and C. Wilson (eds), *Environmental Quality*, Scottish Academic Press, Edinburgh.

Achieving the Correct Balance

The Manchester Second Runway Project

Prof. Dr. Callum Thomas
(Manchester Metropolitan University)

Introduction

The rapid and continuing growth in demand for air travel brings with it significant social and economic benefit as regions develop a network of direct links to the rest of the world. The aviation industry does however create significant global and local environmental damage and adversely affect the lives of large numbers of people living close to airports. In seeking to achieve sustainable development, governments have to make planning decisions for airport expansion based upon the perceived balance that would be achieved between these benefits and costs.

Across Europe are numerous examples of airports that have failed to gain approval for additional growth simply because of the environmental implications that would arise from it. It is no longer sufficient for airports to simply justify development proposals in terms of meeting demand. It is now necessary for them to more explicitly quantify the social and economic benefits that would arise from the proposal. Similarly it is no longer acceptable for developers to make vague promises about environmental protection. The public require very clear, firm and meaningful commitments about the quality of life and environmental they can expect to enjoy in the future if developments are to proceed.

The way in which an airport seeks to achieve planning approval for further development is critical to its own sustainability. This paper examines the way in which Manchester Airport attempted to address the above issues and achieve the correct balance between environment and development in seeking planning approval for its second runway.

1. What is Manchester Airport?

Manchester is the third largest airport in the UK and within the top 20 international airports in the world. Over the past 10 years it has been one of the fastest growing airports in Europe increasing from 4.5 million passengers in 1980 to 17 million in 1999. Currently 95 airlines operate to 175 charter and scheduled destinations world-wide.

Manchester Airport serves a catchment area of 20 million people throughout the North of England. Over 60% of the UK manufacturing industry lies within a 2 hour drive of Manchester and the Airport is seen as one of the most significant generators of economic activity in the region. It could be argued that the future development of the region is intrinsically linked to its network of air routes which provide direct access to the global economy and in this regard, Manchester Airport is currently the single most important provider of such services in the region.

2. Why did Manchester Need a Second Runway?

To cater for increasing passenger demand the Airport Company had embarked upon an extensive development programme, upgrading its existing facilities and providing new passenger terminals, cargo and aircraft maintenance facilities. The Company also began to develop its ground transport links though the construction of a rail station and new coach and bus facilities.

The Airport still only had one runway and already by 1992, this was beginning to be full at peak times. In 1993 the Airport Company reported a peak hour demand of 68 slots although the existing runway facilities could only accommodate 42 movements per hour. In consequence, some airlines were forced to accept what they considered less than ideal slots and some other services failed to materialise.

It was argued therefore that the next constraint upon the growth of the Airport would be runway capacity and that failure to provide sufficient runway slots to meet demand could put at risk the already extensive investment in the terminal and other operational infrastructure. It was also believed that failure to provide additional runway capacity could lead to valuable air routes being lost or jeopardise the development of new routes and services which would have broader regional development implications. At risk was the prospect of other airports in Europe gaining new air services which could have come to the North of England.

The Airport Company maintained that passenger demand would increase to 30 million ppa by the year 2005, levels of growth which are consistent with those being forecast elsewhere in Europe. This reflects an increasing propensity to travel linked to increasing international trade and increasing affluence. At the same time, however increasing affluence leads to increasing expectation of quality of life and less willingness to tolerate disturbance and pollution.

Manchester Airport lies approximately 15 kms to the south of the City of Manchester on the southern edge of the Greater Manchester conurbation in the 'green belt'. Immediately to the north and east of the airport lie extensive residential areas whilst to the south and west lies attractive open countryside. The adjacent area of North Cheshire is one of the most prosperous parts of the North

West, in contrast, Wythenshawe , located close to the northern perimeter of the Airport is one of the largest public housing schemes in Europe and a site of high unemployment. Manchester's second runway project is, therefore a very good example of the difficulties faced in balancing environmental and community protection against the economic and social benefits of development.

Manchester Airport is owned by the City of Manchester and the 9 other councils within Greater Manchester. This ownership provides a very interesting additional facet to the second runway development since some members of the Airport Board are elected by communities which would be adversely affected by the runway whilst others represent constituencies which would only gain benefit from its construction.

3. The Social and Economic Benefits Which Would Arise from the Development

The activities associated with the growth and operation of an airport have the potential to bring significant social and economic benefit to the region serviced by that airport. Studies have suggetsed that major international airports can be the largest generators of economic activity in the regions they serve. These benefits are measured in terms of:

- the creation of direct and indirect employment;
- the facilitation of inward investment;
- trade with the rest of the world;
- travel for education and leisure; and
- the cultural development of the region.

These benefits are however dissipated across the entire region, whereas the costs of airport expansion and operation are borne by the local environmental and communities nearby.

In its submission to the second runway public inquiry, Manchester Airport provided evidence about job creation. Not simply the jobs associated with the construction of the runway but also the longer term employment across the region which would arise from the addition traffic which would be generated by it.

To address concerns about the adverse distribution of the benefits and the impact of the development the Airport Company developed an employment and construction charter with the aim of recruiting a higher proportion of residents of communities adversely affected by its operation.

The Airport Company researched and then provided information about Companies that had moved into the Manchester area because of the existence of a major world airport close by. Support was provided by airlines who illustrated the potential markets from leisure and business which could be found through Manchester Airport and which could not be supported elsewhere.

Finally the role of the Airport Company in supporting regional economic and cultural development was described. This through direct support in inward investment and regeneration organisations, through support for the arts and for initiatives such as the Manchester Olympic Bid which would bring additional opportunity for the people of Northern Britain.

The justification of, and benefits which would accrue from the continued growth of Manchester Airport through the construction of a second runway were, therefore, articulated in considerable detail.

4. Consultation and Environmental Assessment

Manchester Airport's draft development strategy to 2005 introduced the concept of a second runway into the public domain in August 1991. The document set out the Company's long-term air traffic forecasts and its need for additional runway capacity and offered three locations where the runway could be located.

The decision to introduce the runway proposal to the public and statutory bodies responsible for environmental protection at an early stage was a conscious one on the part of the Airport Company. This approach has a number of advantages and disadvantages for the Airport. Statutory bodies would be required to object on principle because of the adverse environmental impact of the proposal. They could, however be very useful in developing mitigation measures designed to offset those effects, if the development were to proceed.

An extensive social survey undertaken prior to the preparation of the development strategy had already provided an insight into the way in which the Airport impacts upon the lives of its neighbours and what action they would like to seen taken to improve their quality of life. The Company also sought public involvement while the planning for the project was in its formative stages otherwise it would have been more difficult to modify the selection process or the design of the proposal to take account of their input.

The disadvantages of this approach are obvious, in that anxiety and uncertainty was raised among local residents, particularly as it was still some 2 years before the planning application was due to be formally submitted. It was also known, however, that there was a desire amongst local residents for more detailed in-

formation on the runway plans, particularly in terms of environmental issues. Within this process there is another unavoidable dilemma. By deciding to consult with the public at an early stage, there is the criticism over the availability of detailed information. By leaving it until later in the process, the reaction would have been that the Airport's mind was already made up and that it is little more than a public relations exercise. For second runway, the public consultation generated a substantial and often hostile response, particularly from local residents within a 5km radius of the Airport and along its departure and arrival routes. However as might be expected in the wider area, the runway proposals received more widespread support.
The main issues raised during consultation included concern about:

- Disturbance caused by aircraft noise
- Road traffic congestion
- Effect on the Green Belt - protection of wildlife and access for walking
- Air quality

What also became clear was that there was considerable local concern about the environmental impact of the Airport's current operation and an understandable fear of the effects of future growth.
During the consultation exercise the Airport Company announced it's intention to carry out a formal environmental impact assessment and that the results and conclusions of this work would be used to determine the runway option that was finally put forward. To ensure that this work was undertaken properly and seen to be addressing the relevant issues advice was sought advice from the UK's Institute of Environmental Assessment which provided the scoping document for the EA.

A total of 16 specialist consultants were appointed address the major impacts associated with the airport's operation including:

- Noise disturbance
- Water quality
- Ecology
- Human health
- Landscape (the visual impact)
- Agriculture
- Ground transportation
- Listed buildings
- Air quality
- Safety and third party risk

At the same time, specialist consultants were commissioned for the engineering design of the runway. The Airport Company were anxious to set up a clear iterative process to ensure that environmental considerations influenced the second runway design right from the very start. It was also used to ensure that the engineering consultants had access to environmental expertise and that the mitigation measures were developed alongside the scheme and not 'bolted on' at a later stage.

After an extensive public consultation exercise involving public meetings and exhibitions, the Airport Board considered the comments made and selected its preferred option for the second runway. The planning proposal was then submitted and a major public inquiry took place between June 1994 to March 1995 and lasted 101 days.

5. Environmental and Community Guarantees

The concerns amongst local residents about the implications of Manchester's growth for their own quality of life clearly had to be addressed by the Airport Company in a meaningful way. Quite simply, having quantified the environmental implications, it would be necessary for the Airport to offer guarantees that if the development were to go ahead local residents would be adequately protected and if no such protection were forthcoming that they would have recourse to the courts.

Accordingly the Airport worked closely with the two planning authorities affected by the development, to explore issues of greatest concern to communities and statutory bodies. Through this approach it developed a comprehensive package of guarantees and environmental safeguards to protect the local environment and the quality of life of local people. These were then incorporated into a legal agreement.

This 'Section 106' legal agreement contained over 100 points of agreement and 33 legally binding guarantees addressing such diverse issues as:

- the ultimate capacity of the Airport
- action to control the impact of aircraft noise
- protection of the green belt
- the development and promotion of ground public transport
- the provision of sound insulation and vortex compensation scheme

6. Approach to Environmental Protection

I will now go on to describe the types of measures taken and commitments given to provide the level of protection which the Company regarded as appropriate.

Disturbance caused by aircraft noise

Aircraft noise is the single most important environmental issue arising from airport operations. Noise is a form of nuisance and therefore associated with an individual's concept of "quality of life". As such, it would be important to involve the Airport's neighbours in the design of the noise control programme. Community involvement was ensured through the work of the Aircraft Monitoring Group, a working party comprising not only pilots and air traffic controllers but also locally elected community representatives and members of amenity groups opposed to the Airports development. The design option chosen for the second runway gave particular attention to noise issues. A decision was made that existing approach paths and departure routes should be followed as far as possible and this factor alone constrained the development significantly.

Manchester Airport has operated a noise control programme for over 20 years, much of it on a voluntary basis. It therefore chose to incorporate the various elements of this programme within the legal
agreement with its planning authorities.

The agreement required the Airport Company to implement a comprehensive noise control programme which was designed to ensure that in the future the noise climate would not deteriorate beyond that measured in 1992/1993 until at least the year 2011. This would be measured in terms of the area of agreed noise contours and the noise levels generated by the loudest aircraft which use the airport in each year. To achieve this, the agreement provided for:

- Targets and detailed action to phase out Chapter 2 aircraft ahead of legislation.
  (Chapter 2 aircraft ceased to be scheduled at night from 31st October 1995).
- Operational penalties for noisy and off track aircraft.
- Noise related operational charges
- Controls on ground operations (engine testing, APU use etc.)
- Preferential runway use
- Improvements to the preferred noise routes and action to ensure aircraft avoid overflying built up areas.
- Restrictions on the types and numbers of aircraft which could operate at night
- The maintenance of a noise and track monitoring system.
- Public access to the system and the regular publication of the results of monitoring.

- Third party auditing of all systems with auditors offering advise as to where improvements can be made.
- The continuing provision of grants foir the installation of Sound Insulation in properties affected by noise.

Protection of the Green Belt

The Airport lies within a protected area of countryside (known as a "green belt") surrounding the Greater Manchester urban conurbation. Accordingly, the Company's approach to ecological protection, mitigation and landscape would be critical to the success of any development proposals. The disruption of the minimum area of land was obviously very important as well as the need to give particular attention to sites of ecological importance.

The whole of the second runway development site covers an area of about 500 hectares, of which 62 hectares (or 12%) would be given over to runways, taxiways and building. Approximately 150 hectares would be required for airfield grassland with a further 300 hectares or 60% of the development site being landscaped and given over to ecological protection work.
Sensitive sites and habitats would be lost, including 0.5 hectares of 300 year old woodland. Obviously this could not be replaced. The Company did, therefore offered a mitigation package, through an ecological landscaping programme which would partially offset such losses. The work includes the purchase and replanting of new land, the restoration of existing landscapes and most exciting of all the translocation of over 2 hectares of semi mature woodland and 5 hectares of natural pasture. This is a comparatively new and experimental technique designed to retain as much of the original habitat as possible rather than establishing new sites from scratch. Similar approaches were taken in regard to the protection of ponds and wildlife.

Public access to the countryside around the Airport was also important so several kilometres of footpaths and bridleways for walkers and horses were designed into the scheme. Creative landscaping would also ensure that natural features would be used where possible to shield these areas from the visual intrusion of the development. Despite extensive investigation of a variety of options it became apparent that it would be necessary for the second runway to cross the River Bollin, a tributary of the River Mersey. The Airport Company faced two dilemmas in this respect, firstly the effect of the development upon the river itself and second the effect upon the river valley. The Company investigated and then rejected a proposal to construct a new five kilometre long valley around the end of the runway. A tunnel would therefore be necessary. It also modified the design to remove a full length parallel taxiway from the proposal. This would reduce the length of any tunnel from 400 metres to 300 metres and bring with it significant ecological benefits. The diameter of the tunnel was increased so as to allow the river to follow its existing natural course under the runway and to change shape over time. An existing weir across the river would be removed and the river bed sculptured to replace it. This would enable fish to move freely up and down the river for the first time in many years and would be an improvement over the existing situation. The tunnel itself would, however still act as a barrier to wildlife moving up and down the river valley. University research was commissioned to see if it would be possible to light the tunnel so as to permit plant growth. This did not prove a feasible option and as a result an ecological corridor was designed around the end of the runway and back into the river valley on the other side. The inside of the tunnel itself was also designed to encourage bats, breeding birds and other 'cave dwelling' species. Under the conditions of the legal agreement the Airport Company committed itself to fund and implement a fifteen year landscape management programme which would be monitored and regularly reviewed.

Ground Transportation and Road Traffic Congestion

It is self evident that an airport has the potential to have a major impact upon its surrounding communities simply as a result of the road traffic associated with passengers using its facilities. Manchester Airport approached this problem by offering to fund improvements to the existing and motorway links, but more importantly to embark upon a major programme of investment designed to improve the public transport infrastructure. Commitments included in the legal agreement include:

- Investment in further developing existing heavy rail links to the Airport
- Promoting and part funding a tramway link to the centre of Manchester.
- The development and promotion of local bus and national coach services.
- The construction of a new Public Transport Terminal
- Initiatives for staff car sharing
- A pro rata reduction in the availability of car parking
- The use of 10% of the Airport's annual marketing budget to promote the use of public transport.
- The appointment of a Ground Transport Manager
- The establishment of a Ground Transport Steering Group including representatives of many interested parties from across the region.

A target was set that at least 25% of trips to and from the Airport should be made by Public Transport by the year 2005.

7. Conclusion

The proposal to construct a second runway at Manchester would bring considerable regional benefits to the North of England. The only reason not to go ahead with the development would be on environmental grounds. Concern for environmental issues has increased enormously in the last ten years and this was the first airport development in the UK to be subject to the recent EC and UK requirements for Environmental Assessment of major projects.

It was clear that given the rapidly changing social and political climate, the way in which environmental issues were handled would be critical to the success or failure of the second runway project.

Key elements of the approach taken included:

- The employment of environmental consultants at the very beginning of the project.
- Commissioning those consultants to assess the environmental impact, to design mitigation measures
  and to collaborate with the engineering consultants to achieve the final design.
- Undertaking an Environmental Assessment which represents current Best Practice.
- Undertaking extensive local consultation with communities affected and enabling their concerns and wishes to be introduced into the design.
- Being open about the environmental implications and providing information as freely as possible.
- Negotiating with statutory bodies responsible for environmental protection so gain their acceptance of the mitigation proposals even though they had to object to the planning application on principle.
- Setting clear and meaningful environmental commitments and performance targets and incorporating these into both a Corporate environment strategy and a voluntary, but legally binding agreement with the Local Planning Authorities.
- Providing environmental monitoring and reporting systems with third party auditing of performance.

This approach was designed to reassure all parties involved that the Company would honour it's environmental commitments relating both to the project itself and also to the 'quality' of the future environment as the airport grew.

It is important when debating major development projects such as the second runway that the regional economic benefits are accurately assessed and highlighted within the case. Similarly, the true extent of the environmental damage that would arise from the development itself and the additional traffic it would create should not be over or understated. Only through this approach is it possible for society to make valued judgements about the acceptability of future growth. This is essential to the concept of sustainability at a local level.

Handling major development issues with the local community is a highly complex and political task for an airport operator. It requires an understanding of the concerns and emotions of residents and the particular environmental issues facing different communities. It is wholly unlike the more usual public relations efforts of the Airport Company in that it is geared up towards developing their involvement in the process and the provision of information, rather than selling the Airport and its services. Indeed overselling the Airport's case and not fully acknowledging the environmental impact can be counter productive.

An airport will never be able to achieve total support for expansion, perhaps not even majority support at a very local level for its future development projects. Perhaps the best that can be hoped is to minimise the opposition through education and the provision of environmental and community guarantees. Above all it is important that airport management understand and indeed accept the genuine concerns and fears of those affected by the development and make every effort to address those concerns in a meaningful and open way.

Hamburg Airport – Environmental Management System in Theory and Practice

Axel Schmidt
(Environmental Affairs Dept.)
Hamburg Airport

Introduction

For years international aviation has been a symbol of a world economy that is coming closer together and of worldwide mobility. However, in the meantime, it has also become a symbol of regional and global environmental problems. The potential global changes currently being discussed in public and described by such catchwords as "greenhouse effect" and "climate", are now also being attributed to international aviation, and regionally to airports as a result of the high air traffic densities they handle. Commercial airports are very complex organisms (requiring a high degree of organisation) on account of their spatial expansion, the wide range of duties undertaken there, their diverse links with customers, contractual partners and authorities, and also as a result of the direct proximity to urban areas. Consequently, the problem of environmental protection represents a major challenge in terms of finance, personnel and organisation for the operators of commercial airports.

In addition to the generally known and much discussed environmental effects of aviation (noise and smell nuisance and air pollutants in the vicinity of airports, suspected contribution to the greenhouse effect and changing the world's climate), there are also other environmental problems which are the result of the spatial size of airports and their intense frequentation. The problems are related to power and water consumption, waste disposal, the heavy traffic load in the vicinity of airports caused by planes, arriving and departing passengers, the large number of people working at airports and by air freight operations. The possibility of major environmental damage due to emergencies and accidents (e.g. fires, crashes, etc.) must also be taken into account when assessing the risks. Moreover, airports are the points of crystallization of widely differing interests.

The technical capabilities to reduce the burden on the environment have been more or less exhausted, but there is still considerable scope on the organizational and management side to significantly reduce the environmental effects on neighbouring areas. A powerful concentrating effect regarding the environmental situation for all parties concerned in a location like the Hamburg Airport can not only be significantly improved by a stringent environment management system that can be clearly defined and implemented, and can also be utilized in the best possible manner.

Existing standards, rules and regulations for the implementation and certification of environment management systems - in this particular case ISO 14001 and the far-reaching EC Environmental Management Audit System (EMAS) - have resulted in the intention to have the structures and activities in the environment area for the Hamburg Airport certified and validitated in keeping with these guidelines.

Theoretical comparison

ISO 14001 (DIN EN ISO 14000ff) is a worldwide valid standard for the creation of an environmental management system as a private sector regulation. Companies from all areas, i.e. also service rendering companies such as the FHG, can apply for certification. Companies particularly from the Anglo-Saxon and Asian areas, as well as companies engaged in worldwide competition, are setting up environmental management systems on the basis of this standard. The "parents" of this standard are the quality management standard ISO 9000ff and the British Standard for Environmental Management Systems BS 7750.

EC Environment Audit, also known as EC Eco-Audit or EMAS (Environmental Management Audit System): This system is only valid within the European Union. It is based on an EC ordinance, thus making it a state system. Originally, only commercial (i.e. manufacturing) companies were able to participate. However, at least in Germany, the system is currently being extended to some services sectors, e.g. in the transport area. In many EU countries the Environment Audit enjoys a higher rating than ISO 14001 as it is more publicly known and assigns greater importance to environmental protection. The principal difference to the ISO standard is the fact that the public must be informed in detail by an Environment Declaration. Moreover, certification must be preceded by an environmental audit, i.e. a report on the environmental effects must be drawn up. Consequently, most authorities refer to EMAS.

The use of environmental technology in Hamburg has been continuous during the past years and it has been particularly innovative with its own ideas in areas that had to be adapted to regional requirements.

The king-pins remain internal and external communication to improve environment protection measures for the immediate area around the airport. The outwardly effective Environment Declaration which, according to EMAS, has to be published in an abbreviated form ever year and in a detailed form every three

years, has the purpose of ensuring effective and public in-situ control of the environmental management system.
The demanded continuous improvement is taken into account by EMAS because the environmental aims (and consequently also the programme) are defined in concrete figures. The aim that is to be achieved, the necessary period of time and the required finances, all become binding.

Transfer to practice

This theoretical approach was backed by an "environmental assessment" conducted at *Flughafen Hamburg GmbH* over a period of 10 years because a practice-oriented system to deal with environmental matters was established in the course of the decade. The technical implementation of political, neighbourhood and authority matters and demands all came together here as a kind of junction covering in a responsible manner the areas of noise protection, maintaining clean air, waste management, water protection, power supply, resources consumption and bird life.

This basis made it possible to adopt environmental guidelines which have long characterized the environmental policy practiced by Flughafen Hamburg GmbH. This proceeded hand-in-hand with the creation of a comprehensive environment manual to regulate the courses adopted in practice for everyone to read, and to define the aims as a guarantee for all parties to concern themselves honestly with environmental matters. This has been undertaken in its entirety by Flughafen Hamburg GmbH.

However, how is the environment management system to be applied to the Hamburg Airport and how can it be made clear that the environment cannot be considered on its own as an isolated matter? How can bold measures can lead to environment-friendly, life-supporting work and global environment thinking and action.

Examples of environmental concern:

- **Noise** (ground noise/flight noise/noise medicine)
  - Measurements, calculations, noise protection programmes and concern for the individual neighbourhoods, public relations work; in the 4th sound protection programme also in conjunction with thermal *insulation*
- **Protection of the waters** (ground water/fire-fighting services/ building sites)
  - Diminishing the winter service burden concerning water, TOC measuring installations.
  - Monitoring waste-water treatment plants/rain water retention basins; enforcing the monitoring functions described in the environment manual

- Network to measure the ground water and judicious resources management.
- At-source reduction of the amount of de-icing agents consumed by adopting a two-step process. First step: De-icing with hot water to which glycol is added. Second step: Application of viscous glycol to prevent re-icing.

- **Air pollution** (establishing PCA use as defined in the airport operation ordinance)
  - Active emission reduction (better engines, turning off the APUs, use of GPUs)
  - Optimization of unit-type district heating power stations > 90% efficiency
  - Passive (filter development for fans and local public traffic)
  - Approval procedures in keeping with the Emission Declaration (BlmSchgG)
  - Linking thermal insulation AGENDA 21 with CO_2 reduction
  - Linking up with the suburban railway system: Aim -> 20% reduction of the emission caused by road traffic

- **Waste** (more intensive sorting of waste/environment-friendly purchases)
  - Planning and organizing waste disposal, commercial waste, on-board waste and building cleaning

- **Labour protection** (improvement in conjunction with environmental protection)
  - Air and noise emissions and immissions
  - Hazardous substances/water-endangering substances

The purpose of these practice-related examples is to show how locally operating airlines, companies, authorities, Lufthansa Technik AG and Flughafen Hamburg GmbH are all engaged in interdisciplinary co-operation to conform with the political and their own environmental aims. This can be indicated by the noise quota which states that future noise development must not exceed the noise base year of 1997 in order to avoid more political restrictions. Interdisciplinary data must be gathered, and the basis for comparative calculations must be created, so that measures can be quickly taken, and in due course, to reduce pollution levels. There must exist a joint willingness to implement this aim on behalf of the environment. The noise quota is also closely linked to housing limitation in the immediate area around the airport.

In the wake of this practical excursion one should remember that an active link to the sciences and research is not only desirable but also necessary to enhance the environmental conditions, both regionally and globally, in order to improve the technical facilities. (I stated at the beginning that the basic technical resources are in place.) Good examples are the many dissertations which are then technically implemented in connection with the airport and which have created a pool of scientific know-how in the course of the years.

The same applies to joint utilization in conjunction with the functions of the established quality management system.

When compared with other transport operators, and in awareness of the many voluntary measures adopted for the Hamburg Airport (and also other international commercial airports), the following factors must be taken into account within a company in the course of implementation for certification according to EMAS or ISO 14001 (provided that a binding environmental report has been published in connection with ISO 14001):

- advance confidence in companies in connection with the implementation of the environmental management systems

- possible reduction of authority conditions, e.g. short-term recurring reporting obligations to the monitoring authorities

- intensified self-monitoring and recognition of internal audits and other preparatory work

- reducing the premiums in connection with environment insurance policies

- simplified credit granting and preferential treatment in this context

- simplified building planning

- legal stipulation for self-substantiation before court

- simple and trusting implementation of a "round table" in a regional sense

- the pervading realization for other disciplines (building/ maintenance, etc.) that environmental protection involves a high measure of individual improvement that not only causes costs but can also result of commercial savings.

Aircraft Engine Emission Charges at Zurich Airport Introduced on September 1st, 1997

Emanuel Fleuti, Head of Environment
(Zurich Airport Authority)

Contents

Purpose and Background
The Model for the Emission Charge
The Implementation of the Charge

Abstract

Based on national legislation and in accordance with national,cantonal and regional clean air policies, Zurich Airport - like other Swiss airports - is introducing an emission charge which is based on engine emission and performance parameters. The aim is to reduce specific emissions from air traffic in order to satisfy political conditions in the context of airport expansion plans required to meet future demands of air traffic. The airport authority believes more in levying emission fees to accelerate the introduction of modern engine technologies than in issuing emission related operating limits. Revenues of the charge are used to finance programs that are related to aircraft emissions.

Following a recommendation of the Federal Department of Transport and Energy, aircraft engines are classified according to their engine emission factor which - for jet engines - is the sum of the LTO-NOx and LTO-HC emissions devided by thrust. For shaft engines, a similar procedure is applied. Depending on the class, a charge of between 0 to 40% of the landing fee will have to be paid to cover expenses related to aircraft emission reductions programs. As a general reduction of the weight-based landing fee of 5% is given for all aircraft, the airport is not increasing its revenue due to this charge.

Purpose and Background

The clean air policy

According to the national clean air concept laid down by the federal parliament, emissions of air pollution substances are to be reduced to the levels of 1960. One of the tools is the Clean Air Ordinance 1983. Air pollution is to be reduced where it exceeds existing limits and the cantons are under obligation to set up a "clean air action plan". The Canton of Zurich did so in setting up its program in April 1990, revising it in June 1996, confirming that all emission sources have to contribute their share to the overall reductions.

Calculations and measuring campaigns showed exceeded values for nitrous dioxide (NO_2) and ozone (O_3) in the region North of Zurich where the airport is located. The airport contributed 6.8% to the NOx emissions in the canton in 1995 - landside car traffic related to the airport not included. Air traffic itself is the largest single source of NOx emissions. As modern jet engines generally have lower emissions except for NOx, special emphasis was given to stabilize the airport related NOx emissions despite expected traffic increase. A number of measures have been proposed by the airport and endorsed by the cantonal government. The most important of these is the proposal for an emission charge. With the canton implementing its mitigation plan, the pollution situation will improve to mostly acceptable values. The airport participating, its share would only increase to 9.1%; if not, it will exceed 10%.

The plans of Zurich Airport

To meet future demands of infrastructure for aviation, Zurich Airport with its partners set up the expansion plan "Airport 2000" involving investments of approximately 2 billion CHF (1.8 billion US$). It includes additionaltaxiways, a new terminal, expansion of cargo facilities and improvements on the landside traffic system, but leaves the runway system unchanged. The environmental impact study - mandatory for such a project - showed an increase in emissions in the "business as usual" scenario. The cantonal government, being the airport operator, then had to confirm its willingness to take appropriate measures to prevent an increase of air pollution in the region.

What are the consequences to bear?

Detailed emission inventories showed air traffic to be the largest single source for NOx. In order to improve at least the specific emissions of aircraft engines, best available technology needs to be used. Therefore, the Canton of Zurich submitted a proposal to the federal authorities to introduce emission-related charges as an economic incentive.

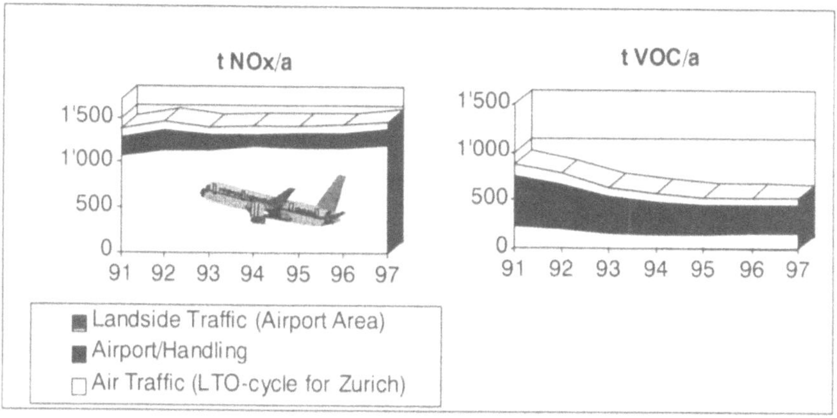

Figure 2: Emissions in the Zurich Airportarea. The air traffic statistics for 1997 show a total of 276,000 movements, 18.3 mio passengers used the airport in 1997. LTO-cycle modified for Zurich (individual taxi time).

The legal aspect

Due to the legal situation and the time frame involved, the authorities agreed on first introducing emission charges in accordance with aviation law, i.e. limited to cover airport emissions reduction measures only, and to postpone an extension to a more effective charge where very low emission aircraft would be subject to a refund.

The revised aviation law became effective on 1 January 1995. The major change concerning environmental protection has been article 39, stating that "when fixing landing charges, airports have to take into account differing noise and gaseous emissions of aircraft". Noise related landing charges in Switzerland have been in use since 1980, whereas emission charges have been unknown so far. According to the revised law, it is within the competence of the airport operator to define the charge, although the regular procedure involves consultations with other airports, airlines and federal authorities.

The concession for the airport expansion program contains the obligation to introduce the emission charge in Zurich.

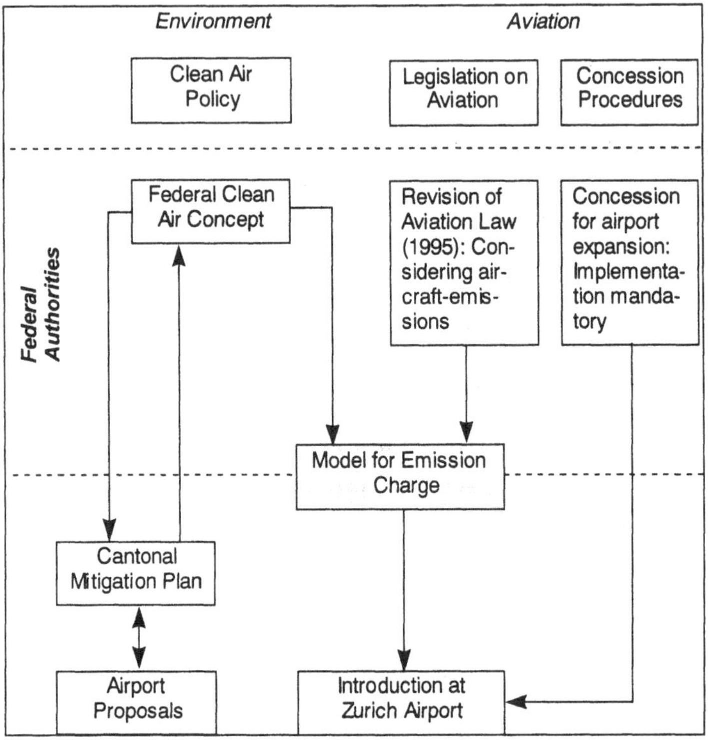

Figure 3: Emission charge based on three legal aspects: Clean air concept, aviation law and concession for expansion program.

What is the purpose of this emission charge?

Primarily, the emission-related charge is there to create incentives: The specific emissions from air traffic are to be reduced by promoting and accelerating the introduction and use of the best available engine technology in order tostabilize airport emissions without having to set limits to operations.

The Federal Office for Civil Aviation (FOCA), in co-operation with the Federal Office for Environmental Protection, Zurich and Geneva Airport Authorities and an airline have worked out a model for the classification of engines according to their emission characteristics and for application of related charges. The main guidelines were to relate charges to aircraft engine emissions, to secure operational freedom of airline operators and to keep the revenue of the charges bound for air traffic emission reduction programs.

The different elements of this emission charge have been laid down in a report which is the basis for a directive of the Federal Department of Transport and Energy which has been passed on to all Swiss airports in August 1996.

The Model for the Emission Charge

Which pollutants are being considered?

To characterize emissions from aircraft engines in this context, it is sufficient to consider the pollutants nitrous oxides (NOx) and hydrocarbons (VOC - volatile organic compounds). These two substances are the main contributors to cumbustion-related air pollution, preceders of ozone, and they are explicitly designated for reduction in the Federal Clean Air Concept. Carbon monoxide (CO) and sulphur dioxide (SO_2) no longer cause a significant pollution problem, whereas other substances are more relevant for the greenhouse effect than as pollutants. CO_2 emission is proportional to fuel consumption. Because fuel efficiency has a very high priority in aircraft engine design by definition, CO_2 need not be taken as a factor for incentive in this context.

What about improved engine technology?

Engine manufacturers have achieved big improvements in fuel efficiency over the last decades. The developments of combustion systems have also brought a decrease of VOC emisson indices but a relative increase in NOx. The new technologies, coming into service now, promise a reduction of both pollutants. Decreasing emission indices and better fuel efficiency lead to an overall emission reduction which should be further promoted.

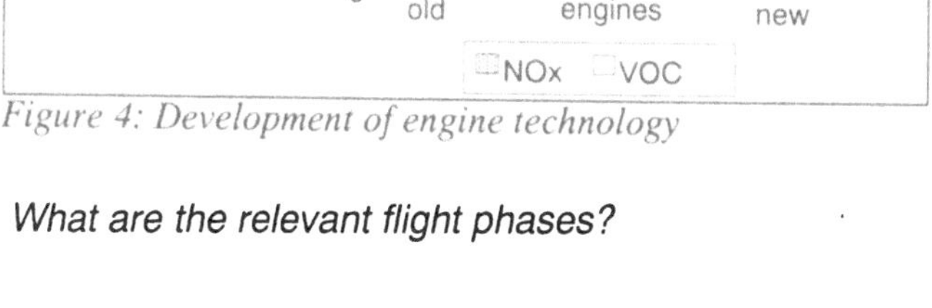

Figure 4: Development of engine technology

What are the relevant flight phases?

When looking at aircraft emissions, the global aspects and therefore emissions during an entire flight should be given due consideration. A standard flight usually covers different height bands.
However, flight profiles are very individual and no internationallystandardized cycle for entire flights exists. A flight cycle has been defined for engine emission purposes. It covers phases within the planetary boundary layer, where procedures and data are accepted and published (LTO cycle, Landing and Take-Off), for jet engines of more than 26.7 kN thrust. The LTO cycle describes 20 to 30% of the NOx-emissions of an entire flight and even up to 90% of the VOC-emissions. Using these LTO cylce emissions for classifying engines will have a positive effect also in reducing emissions at altitude, although not proportionally in every case.

The decision has been taken to use the LTO cycle only as reference for engine emissions.

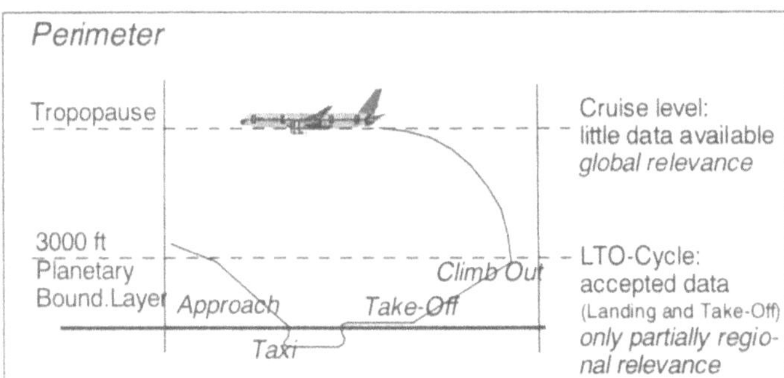

Figure 5: Flight profiles and available data

Is the size of an aircraft taken into account?

In order to classify aircraft, the emissions need to be scaled according to the size of the aircraft: A large aircraft obviously emits more pollutants than a small one, but has also a bigger transport capacity. A possible scale is the maximum take-off weight of an aircraft (MTOW). Another possible scale factor is a characteristics of the engine itself, the maximum installed thrust. Thrust reflects not only the size of the aircraft (bigger aircraft need bigger engines) but also the technical development (increasing thrust by weight ratio).

Where is the emission data taken from?

The characteristic emission data for turbofan engines with more than 26.7 kN thrust is basically obtained from the ICAO document 9646-AN/943 (1995) with its addendae (Engine Exhaust Emission Data Bank).
Data for non-certified engine is collected from sources like the FAA Aircraft Engine Emissions Database (FAEED) or from aircraft engine manufacturers. All data is approved by the FOCA for use by Swiss airports.

What is the calculation procedure?

Based on all the above considerations, the engine emission factor is defined as:

$\mathbf{EEF_{TF} = (NOx_{LTO} + VOC_{LTO}) / max.\ thrust}$	[g/kN] for turbofan engines
$\mathbf{EEF_{TS} = (NOx_{LTO} + VOC_{LTO}) / power}$	[mg/hp] for turboshaft engines

All engines are ranked according to their EEF, returning two lists for classifying.

According to the emission certification data every aircraft operator will be able the calculate the EEF for each aircraft in his fleet.

What happens with engines without data sheet?

So far, only engines with detailed emission data have been addressed. However, for reasons of equal treatment, other engines like piston engines must be considered as well. As long as there are no detailed data available, all these engines are subject to a general surcharge to the landing fee.

In 1995, 28% of all movements were by shaft engines (incl general aviation). Looking at emissions, only 2.6% of NOx and 3.2% of VOC emissions were produced by shaft engines.

The Implementation of the Charge

How does the charge look like?

Based on a number of considerations like clean air incentives, available technologies and alternatives, existing and forecast fleet mix, the FOCA defines five classes each for turbofan and shaft engines, giving a range of EEF within each class.

The definition of the classes is in the responsibility of the FOCA in order to ensure uniform criteria for all Swiss airports.

The charge itself is then expressed as a percentage of the regular landing fee which is added to the landing fee as such and is to be paid by every airline operator. It is within the competence of each airport operator to determine the percentage per emission class, because income requirements are calculated on the basis of the airport operator's medium term budget for costs related to air traffic emissions.
The airport operator may basically change the percentages at any time, but for reasons of transparency and long term planning, they will not be altered too frequently.

The charge shall not be used for financing general costs of infrastructure or as a tax but rather be used for financing costs related to reductions of emissions from air traffic.

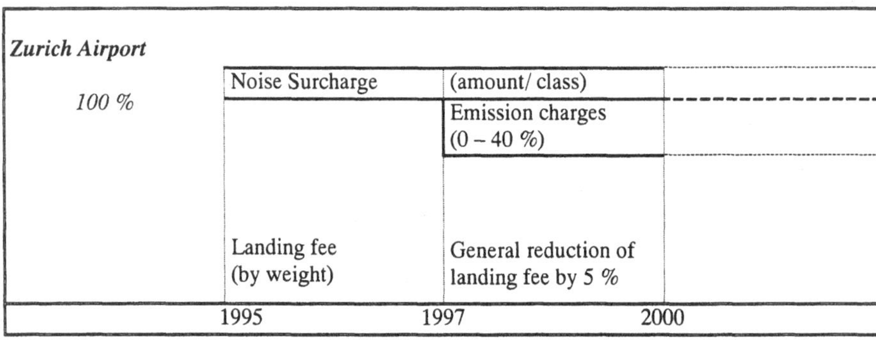

Figure 6: Effect of emission charge on the airport's revenues.

How much has to be paid in Zurich?

The introduction of the emission charge is not just a general increase of the landing fee. Based on the 1996 financial review and a forecast, the existing weight based landing fee has generally been decreased by 5%. Costs for emission reduction measures that have been covered by general finances, will now be financed with the emission charge according to the polluter-pays principle. The total emission charge revenue amounted to approx. 4.5 mio CHF in 1998.

The effect for the airport is such that the overall revenues from air traffic remain unchanged. The emission charge is just compensating the overall reduction in the landing fee.

The effect for the airlines are far less stringent than what might be expected. Based on 1996 traffic figures, 2/3 of all landings are not affected by an increase due to the new charge. By 1998, almost 60% of all movements took place in class 5 (net discount of 5% of landing fee).

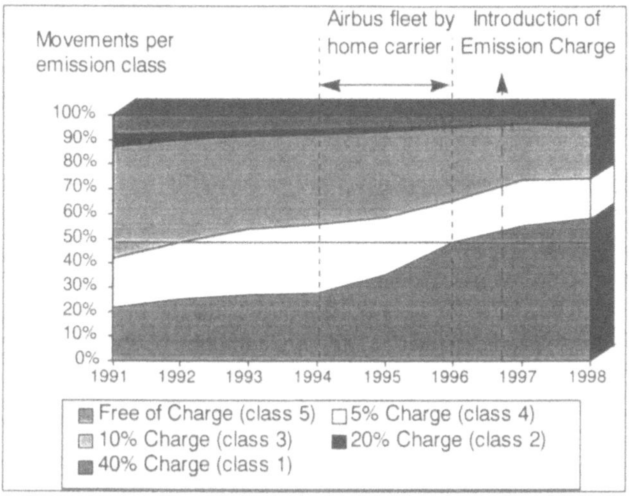

Figure 7: Development of Engine Technology shown by the movements per emission class.

How are the revenues going to be used?

Today's activities as well as the expansion program for Zurich Airport includes a number of projects involving environmental protection related to aircraft emissions. Depending on the ecological benefit of any measure is the share of the costs that are covered by the emission charge.

It varies between 100% (measure entirely for air pollution reasons) and 1% (measure helps to reduce emissions, without being responsible for the benefit). Examples:

- air pollution monitoring station (100%)
- fixed ground power stations (100%)
- additional taxiways to reduce taxi-times (40%)
- aircraft approach/departure system (10%)

The costs amount to approx 4.5 mio CHF/a. Thus, emission charge revenues remain allocated to air traffic purposes.

What happens to excess revenues?

As it is not possible to budget long-term expenditure and income very accurately, unused funds or a shortage of a given year will be considered in the calculation of the following years. Any alteration of the emission charge will take the previous surplus or deficit into account.

Will this charge ever be withdrawn again?

This charge may be withdrawn, when the clean air objectives are achieved (emissions from air traffic are reduced to and stabilized at the level required by the authorities) or if the whole aircraft fleet corresponds with the best available technique. It is likely, however, that a charge will be maintained beyond existing emission reducing costs, but then the funds raised will be redirected to operation of low emission engines.

What is the legal consideration of the model and its implementation?

The proposed classification model and charges are in compliance with Swiss legislation and conform to the CARFM report (Conference on Airport and Route Facility Management, June 1991) as well as to the principle guidelines of ICAO for charges or taxes.

Is it really necessary?

Zurich Airport Authority is run by the Canton of Zurich. Therefore, the credit for the required expansion program had to be approved by a public vote. During the debate, the cantonal parliament decided to raise the issue of an emission charge and made it a condition for taking specific elements of the planned infrastructure into service. Furthermore, Zurich Airport has always been willing to take responsibility in questions of environmental protection.

Independent of economic effects, emission charges may also signal an innovative environmental policy at local or even national level and does improve acceptance of an airport. If such a charge is based on a transparent model and if it is easy to handle, it will likely be introduced at other airports, thus multiplying its benefits.

Biographical Notes

Dr. Carl A.M. Brenninkmeijer

Carl A.M. Brenninkmeijer was born in Amsterdam and educated there as well as in Groningen where he obtained his M.Sc. in experimental physics. Subsequently he received a scholarship for environmental research at the *Institute of Nuclear Sciences* in Lower Hut, New Zealand. After two years of scientific work on problems related to geothermal energy, he returned to Groningen for research towards a Ph.D degree.
After having obtained this degree in 1983 based on his work in isotope palaeoclimatology, he moved with his wife and three children once again to New Zealand, where he became head of the mass spectrometry laboratory of the *Institute of Geological and Nuclear Sciences*. After a period of developing equipment and automating several operations, he turned his attention to atmospheric chemistry with emphasis on the use of isotope anaylses.
After having worked for two years with the *National Institute of Water and Atmospheric Research*, where he coordinated research flights to the Antarctic, he moved to Germany in 1994 together with his family to take up a position as group leader at the *Atmospheric Chemistry Division* of the *Max Planck Institute of Chemistry* in Mainz in order to be fully devoted atomspheric chemistry.

Emanuel Fleuti

Emanuel Fleuti was born in 1962 and studied geography at Bern University until 1990 where he graduated as a qualified geographer (Dipl. Geograph).
In 1990, Mr. Fleuti started his career at Zurich Airport *as Engineer for Environmental Affairs*, reporting to the board of directors.
Meanwhile, the area has formed into a separate department with a team of three people and with Mr. Fleuti in charge. The department is responsible for all environmental issues, except noise abatement which is dealt with by a separate division. Mr. Fleuti‘s special field of occupation is air pollution control.

He also works with ADV and ACI and currently holds the BAZL (Federal Authority for Civil Aviation) mandate in the ECAC.

David W. Gillen

David W. Gillen is Adjunct Professor of *Civil and Environmental Engineering and Research Economist Institute for Transportation Studies* at the University of California, Berkeley. He also holds a position of Professor of Business in Waterloo, Canada. He graduated with a Ph.D. in Economics 1975 from the University of Toronto specializing in Transportation and Industrial Organization. He has taught at universities in Canada and the United States. In 1991, he was appointed as a Senior Advisor to the *Royal Commission on National Passenger Transportation* and led the research on pricing, investment and cost recovery.

Mr. Gillen has published numerous articles, reports and books in the field of Transportation Economics, Industrial Organization and Management Economics. His primary interest is aviation economics and management. He has spent the last decade examining and modelling the strategic behaviour of airlines in their changing regulatory environment. More recently he has shifted his focus to airport management where he has been undertaking research on modern business applications to airport management, performance measurement and productivity assessments.

Peter S. Morrell

Peter Morrell graduated in Economics from Cambridge University and subsequently gained a Masters in air transportation from the Massachusetts Institute of Technology, where he worked on NASA sponsored research into airline forecasting and profitability. He initially worked with merchant bank, Lazard Brothers, in the City, before joining the*Association of European Airlines* in Brussels as an economist in 1971. He worked as an air transport consultant from 1978 to 1991, initially with*Alistair Tucker Associates* and subsequently independently.
He has been involved in numerous airline and airport projects world-wide. Airport consultancy has included work on traffic and financial forecasting and/or privatisation for Hannover, London Luton, Warsaw, Berlin, Manchester and Cyprus airports in Europe, and airports in other parts of the world including Argentina, Mongolia, Brunei, Bangkok and Kathmandu. He recently directed a study for the *European Commission* on the effectiveness of EU measures in bringing about an internal market for air transport. He co-authored a report on the aircraft industry and the potential for emission reductions, which formed part of a study for the *Department of the Environment on Allocation of Emissions from International Bunker Fuels under UN Framework Convention on Climate*

Change. He is also supervising research into the social welfare impact of environmental taxation on air transport.
He has lectured on numerous industry courses, and directs the Airline Finance for Non-Specialists course at Cranfield. He lectures on airline industry regulation for the *Royal Aeronautical Society's* annual Oxford Air Transport Course, has taught airline management at the *French L'Ecole National de l'Aviation Civil* (ENAC), and has lectured on airport project appraisal at the *Brazilian Instituto de Aviaçao Civil*. He has also lectured on in-house management courses for the Estonian airports, airlines which included British Airways, Singapore Airlines, Malaysia Airlines, South African Airways, Finnair, LOT Polish Airlines, Emirates Airlines and Cyprus Airways, and courses for other clients such as British Telecom and Union Bank of Switzerland.
He is now Head of the *Air Transport Group* at the College of Aeronautics at Cranfield University, lecturing there on air transport economics and finance. He also heads the Group's consultancy practice, and was until recently joint editor of the Journal of Air Transport Management.

Dr. Hans-Martin Niemeier

Hans-Martin Niemeier was born in 1956. In 1982, he graduated at Hamburg University with an MA in Economics and remained at university as assistant of Prof. Harald Scherf at the *Institute for Statistics of the University of Hamburg* until 1986.
From 1988-1989, Mr. Niemeier held a position as visiting adjunct at Houston Baptist University lecturing on the *History of Economic Thought and on International Economics*. In 1989, he returned to Germany where in the same year he attained his Ph.D. in Economics at Hamburg University.
Since 1990, Mr. Niemeier has been working for the Senat of the Free Hanseatic City of Hamburg in the aviation section of the *State Ministry of Economic Affairs*.
He directed econometric studies by consultants on the traffic demand of Hamburg International Airport and has also developed concepts for a sustainable economic and environmental development of the airport, such as*Guidelines for a Northern German Aviation Policy, Noise Budget for Hamburg Airport*, etc.

Axel Schmidt

Born in 1958, **Axel Schmidt** grew up in Schleswig-Holstein.
After having finished school and then an apprenticeship, he studied Environmental Technology at Hamburg University from 1980 until 1984. His studies were followed by a two-year-employment in 1984 at a civil and structural engineering office. In 1986, he then worked for the *Community District Environmental Department* in the field of technical conservation and health care where remained for three years.
Since 1989, Mr. Schmidt has been responsible for the *Environmental Unit* at Hamburg Airport as Centre Manager.
He is also the representative of Hamburg Airport's *Noise Abatement Commission*, is an active member of the *ADV* (study group of Germany's public airports), as well as in numerous activities in different working pools. Mr. Schmidt is also a member of the *ACI-Environmental Committee*.

Callum Thomas

Callum Thomas is an internationally known expert working in the field of aviation and environment. Currently he is the Chair of *Sustainable Aviation* at the *Manchester Metropolitan University*. Apart from that he also acts as advisor to the UK Government and European Commission. Mr. Thomas is responsible for establishing and managing *Manchester Airport's Environmental Consultancy Unit*, as well as for formulating and promoting a corporate strategy. This strategy was designed to ensure the future development of Manchester Airport by achieving the correct balance between environmental and community objectives and commercial enterprise.
In 1992, he established the UK's first dedicated Environment Department, a unit which has since won international acclaim for its innovative approach to environmental management. Prior to that Mr. Thomas was responsible for creating and managing Manchester Airport's *Bird Control Unit*, the first of its kind in UK civil aviation. He has a strong academic grounding in research, teaching and publishing, by training a zoologist with a PhD in aviation ecology.

Dr. Upali K. Wickrama

Upali K. Wickrama is head of *Forecasting and Economic Planning of the International Civil Aviation Organization*. In this position, he is responsible for forecasting, economic planning and other economic analysis activities of ICAO, including the economic aspects of CNS/ ATM. He also acts as the Secretary of *ICAO's Pacific Area Traffic Forecasting Group* (APA TFG).
He has been involved extensively in analytical activities of*ICAO's Committee on Aviation Environmental Protection* (CAEP) and is Rapporteur of its *Forecasting and Economic Analysis Support Group* (FESG). Prior to joining ICAO, he worked for the Boeing Commercial Airplane Group in Seattle. During his seventeen years at Boeing, he held several key positions in sales and marketing with responsibilities for forecasting, planning and financial analysis. He has travelled extensively throughout the world for consultation purposes with several airlines. Dr. Wickrama also lectured at the *Seattle International University* in price theory and aconometrics in the MBA program.

For the avoidance of doubt, input of the whole or any part of this publication
into any generative artificial intelligence model, chatbot, program, software or technology
is expressly prohibited without the prior written consent of the publisher.

Printed by
CPI books GmbH, Leck

FSC
www.fsc.org
MIX
Papier | Fördert
gute Waldnutzung
FSC® C083411

Kolibri 360 GmbH
Ferdinand-Jühlke-Straße 7
99095 Erfurt, Deutschland
produktsicherheit@kolibri360.de